FE Exam Review

Electrical and Computer Engineering

Myron E. Sveum, PE, MSEE

Professional Publications, Inc. • Belmont, CA

How to Locate Errata and Other Updates for This Book

At Professional Publications, we do our best to bring you error-free books. But when errors do occur, we want to make sure that you know about them so they cause as little confusion as possible.

A current list of known errata and other updates for this book is available on the PPI website at **www.ppi2pass.com/errata**. We update the errata page as often as necessary, so check in regularly. You will also find instructions for submitting suspected errata. We are grateful to every reader who takes the time to help us improve the quality of our books by pointing out an error.

FE Exam Review: Electrical and Computer Engineering

Current printing of this edition: 1

Printing History

edition number	printing number	update
1	1	New book.

Printed in the United States of America

PPI
1250 Fifth Avenue, Belmont, CA 94002
(650) 593-9119
www.ppi2pass.com

Library of Congress Cataloging-in-Publication Data
Sveum, Myron E.
    FE exam review : electrical and computer engineering / Myron E. Sveum.
        p. cm.
    Includes bibliographical references and index.
    ISBN-13: 978-1-59126-069-1
    ISBN-10: 1-59126-069-8
     1. Electric engineering--United States--Examinations, questions, etc. 2. Electric engineering--Problems, exercises, etc. 3. Engineers--Certification--United States. I. Title.

TK169.S84 2006
621.3076--dc22
                                                            2006045345

Table of Contents

Preface

FE Exam Review: Electrical and Computer Engineering has a twofold purpose: to be a general introduction to electrical and electronic engineering for non-electrical engineering (non-EE) and non-electrical engineering technology (non-EET) students, and to prepare these audiences for the National Council of Examiners for Engineering and Surveying's Fundamentals of Engineering (FE) examination. Students can use this book for self-study, and instructors can use it as a template for course development.

The need for this book quickly became apparent in my early days of teaching FE exam classes. Most non-EE and non-EET students did not have a sufficient understanding of electrical fundamentals to keep up with the lecture. Writing formulas from freshman circuits classes on the whiteboard did not solve the problem. Based on student feedback, I eliminated excessive theory from my lectures and targeted the most relevant topics covered on the FE exam. As soon as I created PowerPoint® slides for the FE exam review and regular EET classes, the lectures went much better on both sides of the lectern. *FE Exam Review: Electrical and Computer Engineering* is the result of putting those lectures and slides into book form. While this book goes somewhat beyond the information needed for the FE exam, I've found that students appreciate the added confidence the comprehensive coverage provides on exam day.

The book chapters cover direct-current circuits, alternating-current circuits, three-phase power and electric machines, operational amplifier and diode circuits, computer hardware, software, and instrumentation. The material will be useful to most in the non-EE and non-EET fields during their school years, on the FE exam, and throughout their careers.

Please visit Professional Publications' website at **www.ppi2pass.com/fefaqs.html** to find the latest advice and information for the FE exam.

Myron E. Sveum, PE, MSEE

Acknowledgments

I thank the students in my FE exam review and regular EET classes for being the inspiration and reason for this book. Subsequently, they were the "guinea pigs" of my class content experimentation, and their patience is greatly appreciated.

I express my great appreciation to the electrical engineers Gregg Wagener, PE, and John A. Camara, PE, who reviewed the material for technical accuracy and appropriateness. Both have contributed their expertise to the U.S. Navy for many years, and I was happy my book could also benefit from the experiences of these men.

I want to acknowledge the PPI staff who helped me bring this book to light: Sean Sullivan, the project editor who worked with me throughout the production of the book and who tracked me in my travels from Alaska to Mexico; Dennis Rowcliffe, the second project editor, who carried it the last few yards after Sean handed it off; Miriam Hanes, who laid the manuscript into finished pages; Tom Bergstrom, who put the polish on my illustrations; and Amy Schwertman, who designed the cover. Finally, I'd like to thank PPI Director of Editorial Sarah Hubbard, and Director of Production Cathy Schrott.

Myron E. Sveum, PE, MSEE

About the FE Examination

The Fundamentals of Engineering (FE) Examination, also known as the Engineer-In-Training (EIT) Examination and the Engineer Intern (EI) Exam, covers basic subjects from the mathematics, physics, chemistry, biology, and engineering classes of an undergraduate engineering curriculum. The National Council of Examiners for Engineering and Surveying (NCEES) produces, distributes, and scores the national exam. Individual states purchase the exams from NCEES, administer the exams and appeals, and notify examinees of the results.

FE EXAMINATION FORMAT

The FE examination has the following format and characteristics.

- There are two four-hour sessions separated by a one-hour lunch period.

- Examination questions are distributed in a bound examination booklet. A different examination booklet is used for each of these two sessions.

- The morning session (also known as the *A.M. session*) has 120 multiple-choice questions, each with four possible answers lettered (A) to (D). Each problem in the morning session is worth one point. The total score possible in the morning is 120 points. Guessing is valid; no points are subtracted for incorrect answers.

There are questions on the morning session examination from most of the undergraduate engineering degree program subjects. Questions from the same subject are all grouped together, and the subjects are labeled. The percentages of questions on each subject in the morning session are given in the following table.

- There are seven different versions of the afternoon session (also known as the *P.M. session*). Six correspond to a specific engineering discipline: chemical, civil, electrical, environmental, industrial, and mechanical engineering. The seventh version of the afternoon examination is a general examination suitable for anyone, but in particular, for engineers whose specialties are not one of the other six disciplines.

Each version of the afternoon session consists of 60 questions that count two points each, for a total

Morning FE Exam Subjects

subject	percentage of total questions (%)
mathematics	15
engineering probability and statistics	7
chemistry	9
computers	7
ethics and business practices	7
engineering economics	8
engineering mechanics (statics and dynamics)	10
strength of materials	7
material properties	7
fluid mechanics	7
electricity and magnetism	9
thermodynamics	7

of 120 points. All questions are mandatory. Each question consists of a problem statement followed by multiple-choice questions. Four answer choices lettered (A) through (D) are given, from which you must choose the best answer. Questions in each subject may be grouped into related problem sets containing between two and ten questions each.

Though the subjects in the general afternoon examination correspond to the morning subjects, the questions are more complex—hence their double weighting. Questions on the afternoon examination are intended to cover concepts learned in the last two years of a four-year degree program. Unlike morning questions, these questions may deal with more than one basic concept per question.

The percentages of questions on each subject in the general afternoon session examination are given in the following table. (This book assumes you will elect to take the general examination.)

The percentages of questions on each subject in the discipline-specific afternoon session examination are listed at the end of this section. The discipline-specific afternoon examinations cover substantially different bodies of knowledge than the morning examination. Formulas and tables of data needed to solve questions in these examinations will be included in either the NCEES Handbook or in the body of the question statement itself.

Afternoon FE Exam Subjects
(General Exam)

subject	percentage of total questions (%)
advanced engineering mathematics	10
engineering probability and statistics	9
biology	5
engineering economics	10
application of engineering mechanics	13
engineering of materials	11
fluids	15
electricity and magnetism	12
thermodynamics and heat transfer	15

- The scores from the morning and afternoon sessions are added together to determine your total score. No points are subtracted for guessing or incorrect answers. Both sessions are given equal weight. It is not necessary to achieve any minimum score on either the morning or afternoon sessions.

- All grading is done by computer optical sensing. Responses must be recorded on special answer sheets with the mechanical pencils provided to examinees by NCEES.

Use of SI Units on the FE Exam

Metric, or SI, units are used in virtually all subjects, except some civil engineering and surveying subjects that typically use only customary U.S. (i.e., English) units.

It is the goal of NCEES to use SI units that are consistent with ANSI/IEEE standard 268-1992 (the American Standard for Metric Practice). Non-SI metric units might still be used when common or where needed for consistency with tabulated data (e.g., use of bars in pressure measurement).

What Does *Most Nearly* Really Mean?

One of the more disquieting aspects of FE exam questions is that the available answer choices are seldom exact. Answer choices generally have only two or three significant digits. Exam questions instruct you to complete the sentence, "The value is most nearly...", or they ask "Which answer choice is closest to the correct value?" A lot of self-confidence is required to move on to the next question when you don't find an exact match for the answer you calculated, or if you have had to split the difference because no available answer choice is close.

The NCEES has described it like this: "Many of the questions on NCEES exams require calculations to arrive at a numerical answer. Depending on the method of calculation used, it is very possible that examinees working correctly will arrive at a range of answers. The phrase 'most nearly' is used to accommodate all these answers that have been derived correctly but which may be slightly different from the correct answer choice given on the exam. You should use good engineering judgment when selecting your choice of answer. For example, if the question asks you to calculate an electrical current or determine the load on a beam, you should literally select the answer option that is most nearly what you calculated, regardless of whether it is more or less than your calculated value. However, if the question asks you to select a fuse or circuit breaker to protect against a calculated current or to size a beam to carry a load, you should select an answer option that will safely carry the current or load. Typically, this requires selecting a value that is closest to but larger than the current or load."

The difference is significant. Suppose you were asked to calculate "most nearly" the volumetric pure water flow required to dilute a contaminated stream to an acceptable concentration. Suppose, also, that you calculated 823 gpm. If the answer choices were (A) 600 gpm, (B) 800 gpm, (C) 1000 gpm, and (D) 1200 gpm, you would go with answer choice (B), because it is most nearly what you calculated. If, however, you were asked to select a pump or pipe with the same rated capacities, you would have to go with choice (C). Got it? If not, stop reading until you understand the distinction.

GRADING AND SCORING THE FE EXAM

The FE exam is not graded on a curve, and there is no guarantee that a certain percent of examinees will pass. Rather, NCEES uses a modification of the Angoff procedure to determine the suggested passing score (the cutoff point or cut score).

With this method, a group of engineering professors and other experts estimate the fraction of minimally qualified engineers that will be able to answer each question correctly. The summation of the estimated fractions for all test questions becomes the passing score. The passing score in recent years has been somewhat less than 50% (i.e., a raw score of approximately 110 points out of 240). Because the law in most states requires engineers to achieve a score of 70% to become licensed, you may be reported as having achieved a score of 70% if your raw score is greater than the passing score established by NCEES, regardless of the raw percentage. The actual score may be slightly more or slightly less than 110 as determined from the performance of all examinees on the equating subtest.

Approximately 20% of each FE exam consists of questions repeated from previous examinations—this is the *equating subtest*. Since the performance of previous examinees on the equating subtest is known, comparisons can be made between the two examinations and examinee populations. These comparisons are used to adjust the passing score.

The individual states are free to adopt their own passing score, but all adopt NCEES's suggested passing score because the states believe this cutoff score can be defended if challenged.

You will receive the results of your examination from your state board (not NCEES) by mail. Allow at least four months for notification. Candidates will receive a pass or fail notice only, and will no longer receive a numerical score. A diagnostic report is provided to those who fail.

See **www.ppi2pass.com/fepassrates.html** for recent and historic FE pass rates.

The following table lists the approximate percentages of examinees passing the FE exam.

Approximate FE Exam Passing Rates

examination module	first-time takers (%)	repeat takers (%)
chemical	84	34
civil	68	16
electrical	67	19
environmental	76	24
industrial	66	16
mechanical	78	22
general	67	14

PERMITTED MATERIALS

Since October 1993, the FE examination has been what NCEES calls a "limited-reference" exam. This means that no books or references other than those supplied by NCEES may be used. Therefore, the FE examination is really an "NCEES-publication only" exam. NCEES provides its own handbook for use during the examination. No books from other publishers may be used.

Calculators

The exam requires use of a scientific calculator. However, it may not be obvious that you should bring a spare calculator with you to the examination. It is always unfortunate when an examinee is not able to finish because his or her calculator was dropped or stolen or stopped working for some unknown reason.

NCEES has banned communicating and text-editing calculators from the exam site. Only select types of calculators are permitted. Check the current list of permissible devices at the Professional Publications website (**www.ppi2pass.com/calculators**). Nomographs and specialty slide rules are permitted.

The exam has not been optimized for any particular brand or type of calculator. In fact, for most calculations, a \$15 scientific calculator will produce results as satisfactory as those from a \$200 calculator. There are definite benefits to having built-in statistical functions, graphing, unit-conversion, and equation-solving capabilities. However, these benefits are not so great as to give anyone an unfair advantage.

It is essential that a calculator used for the civil PE examination have the following functions.

- trigonometric and inverse trigonometric functions

- hyperbolic and inverse hyperbolic functions

- π

- $\sqrt{x}$ and x^2

- both common and natural logarithms

- y^x and e^x

For maximum speed, your calculator should also have or be programmed for the following functions.

- interpolation

- finding standard deviations and variances

- extracting roots of quadratic and higher-order equations

- calculating determinants of matrices

- linear regression

- calculating factors for economic analysis questions

You may not share calculators with other examinees.

Laptop computers are generally not permitted in the examination. Their use has been considered, and some states may actually permit them. However, considering the nature of the exam questions, it is very unlikely that laptops would provide any advantage.

You may not use a walkie-talkie, cell phone, or other communications device during the exam.

Be sure to take your calculator with you whenever you leave the examination room for any length of time.

STRATEGIES FOR PASSING THE FE EXAM

The most successful strategy to pass the FE exam is to prepare in all of the examination subjects. Do not limit the number of subjects you study in hopes of finding enough questions in your particular areas of knowledge to pass.

Fast recall and stamina are essential to doing well. You must be able to quickly recall solution procedures, formulas, and important data. You will not have time during the exam to derive solutions methods—you must know them instinctively. This ability must be maintained for eight hours. If you are using this book to prepare for the FE examination, the best way to develop fast recall and stamina is to work the practice problems at the end of the chapters. Be sure to gain familiarity with the NCEES Handbook by using it as your only reference for most of the problems you work.

In order to get exposure to all examination subjects, it is imperative that you develop and adhere to a review schedule. If you are not taking a classroom review course (where the order of your preparation is determined by the lectures), prepare your own review schedule. For example, plan on covering this book at the rate of one chapter per day in order to finish before the examination date.

There are also physical demands on your body during the examination. It is very difficult to remain tense, alert, and attentive for eight hours or more. Unfortunately, the more time you study, the less time you have to maintain your physical condition. Thus, most examinees arrive at the examination site in peak mental condition but in deteriorated physical condition. While preparing for the FE exam is not the only good reason for embarking on a physical conditioning program, it can serve as a good incentive to get in shape.

It will be helpful to make a few simple decisions prior to starting your review. You should be aware of the different options available to you. For example, you should decide early on to

- use SI units in your preparation

- perform electrical calculations with effective (rms) or maximum values

- take calculations out to a maximum of four significant digits

- prepare in all examination subjects, not just your specialty areas

At the beginning of your review program, you should locate a spare calculator. It is not necessary to buy a spare if you can arrange to borrow one from a friend or the office. However, if possible, your primary and spare

calculators should be identical. If your spare calculator is not identical to the primary calculator, spend a few minutes familiarizing yourself with its functions.

A Few Days Before the Exam

There are a few things you should do a week or so before the examination date. For example, visit the exam site in order to find the building, parking areas, examination room, and rest rooms. You should also make arrangements for child care and transportation. Since the examination does not always start or end at the designated times, make sure that your child care and transportation arrangements can tolerate a later-than-usual completion.

Second in importance to your scholastic preparation is the preparation of your two examination kits. The first kit consists of a bag or box containing items to bring with you into the examination room. NCEES provides mechanical pencils for use in the exam. It is not necessary (nor is it permitted) for you to bring your own pencils or erasers.

[ ]   letter admitting you to the examination
[ ]   photographic identification
[ ]   main calculator
[ ]   spare calculator
[ ]   extra calculator batteries
[ ]   unobtrusive snacks
[ ]   travel pack of tissues
[ ]   headache remedy
[ ] $2.00 in change
[ ]   light, comfortable sweater
[ ]   loose shoes or slippers
[ ]   cushion for your chair
[ ]   small hand towel
[ ]   earplugs
[ ]   wristwatch with alarm
[ ]   wire coat hanger
[ ]   extra set of car keys

The second kit consists of the following items and should be left in a separate bag or box in your car in case they are needed.

[ ]   copy of your application
[ ]   proof of delivery
[ ]   this book
[ ]   other references
[ ]   regular dictionary
[ ]   scientific dictionary
[ ]   course notes in three-ring binders
[ ]   cardboard box (use as a bookcase)
[ ]   instruction booklets for all your calculators
[ ]   light lunch
[ ]   beverages in thermos and cans
[ ]   sunglasses

[ ]   extra pair of prescription glasses
[ ]   raincoat, boots, gloves, hat, and umbrella
[ ]   street map of the examination site
[ ]   note to the parking patrol for your windshield
[ ]   battery powered desk lamp

The Day Before the Exam

Take off the day before the examination from work to relax. Do not cram the last night. A good prior night's sleep is the best way to start the examination. If you live far from the examination site, consider getting a hotel room in which to spend the night.

Make sure your exam kits are packed and ready to go.

The Day of the Exam

You should arrive at least 30 minutes before the examination starts. This will allow time for finding a convenient parking place, bringing your materials to the examination room, and making room and seating changes. Be prepared, though, to find that the examination room is not open or ready at the designated time.

Once the examination has started, observe the following suggestions.

- Set your wristwatch alarm for five minutes before the end of each four-hour session and use that remaining time to guess at all of the remaining unsolved problems. Do not work up until the very end. You will be successful with about 25 percent of your guesses, and these points will more than make up for the few points you might earn by working during the last five minutes.

- Do not spend more than two minutes per morning question. (The average time available per problem is two minutes.) If you have not finished a question in that time, make a note of it and continue on.

- Do not spend time trying to ask your proctors technical questions. Even if they are knowledgeable in engineering, they will not be permitted to answer your questions.

- Make a quick mental note about any problems for which you cannot find a correct response or for which you believe there are two correct answers. Errors in the exam are rare, but they do occur. Being able to point out an error later might give you the margin you need to pass. Since such problems are almost always discovered during the scoring process and discounted from the examination, it is not necessary to tell your proctor, but be sure to mark the one best answer before moving on.

- Make sure all of your responses on the answer sheet are dark and completely fill the bubbles.

How to Use This Book

HOW EXAMINEES CAN USE THIS BOOK

FE Exam Review: Electrical and Computer Engineering is designed to get you ready for the electrical and computer subjects on the FE examination. This book is not intended as a reference book, because it cannot be used in the FE exam. On the other hand, the book that may be used during the exam (the NCEES Handbook) is not suitable for use as a study aid. However, it is a good idea to become familiar with the handbook's format, layout, and organization.

For best results, this book should be used in a manner similar to the following outline.

- Study every subject in this book. NCEES has already greatly reduced the scope of the FE exam and simplified the problems. The main thing is how fast you can work problems outside of your favorite subjects. Familiarity with this book's electrical subjects and the related NCEES Handbook content will give you speed on these types of problems on the exam.

- Establish a study schedule to plan your approach. Use days off to rest, review, and study problems from other books.

- Obtain a copy of the NCEES Handbook.* Use it to solve most of the practice problems in this book. By the end of your review, you should know the order of the chapters, what data is included, and the approximate locations of important figures and tables.

- Solve every FE-style practice problem in this book. Don't skip any of them. The problems in this book were included for a reason. Don't short circuit your review by skipping problems that you know are important. The solutions have been isolated at the end of the chapter to help you solve the problems under realistic conditions. Refer to the reference material in the chapter. The chapter's explanation are not in the NCEES Handbook, but the formulas, tables, and figures are all similar.

- Review each chapter's text before tackling its problems. Unlike reference books that you only refer to when needed, you need to read each chapter. That's going to be your only review. Most of the information has a high probability of showing up on the exam.

- Review the example problems in the chapters. There are a number of these in each chapter. Either solve them (and cover up the solutions), or just read through them and check that you know what is going on. It is easier to evaluate whether or not you know the material if you actually do the problem.

- The FE exam primarily uses SI units, so the need to work problems in two unit systems is greatly diminished. Learn the SI system if you are not already familiar with it.

The following list highlights important information for those taking the FE exam. The subjects are covered in more detail in the chapters. Exam candidates should know how to do these tasks.

- Find power dissipated in a resistor or similar device in a DC circuit. Use $P = VI$ and its variations (Ch. 1).

- Find energy dissipated in a resistor or similar device in a DC circuit. Use $E = Pt$ (Ch. 1).

- Use superposition analysis to simplify DC circuits that use two or more sources (Ch. 1).

- Find the Thevenin and Norton equivalents (Ch. 1).

- Know that Norton equivalent resistance is always the same as Thevenin equivalent resistance for any particular DC circuit, a fact that provides a check on circuit analysis (Ch. 1).

- Recognize that a Norton equivalent and a Thevenin equivalent will appear identical to measurements made at the output terminals of a DC circuit: There is no way to distinguish them from this perspective (Ch. 1).

- Remember that maximum power is transferred from a DC power source to the load when the Norton or Thevenin equivalent resistance equals the load resistance (Ch. 1).

---

*If you do not receive a copy of this from your state engineering licensing board, you can obtain a copy from PPI.

- Know that the beginning of the trace in AC waveform analysis is not necessarily the start of the period (Ch. 2).

- Avoid the mistake of applying the rms formula to a non-sinusoidal waveform. The formula can only be used for sine waves (Ch. 2).

- Distinguish differences and similarities in comparing capacitors and inductors to resistors (Ch. 2).

- Know that a transformer does not create power, and an ideal transformer does not lose power (Ch. 2).

- Remember that in three-phase power problems, $\cos \theta$ is the power factor, not the phase angle between the legs (Ch. 3).

- Understand that a transformer must be driven by AC, not by DC or a non-changing input (Ch. 2).

- Read a short subprogram in structured programming statements and determine the result. These statements are usually in pseudocode that does not require expertise in any one language (Ch. 6).

- Know the differences and characteristics of single phase power and three phase power systems, and why three phase power systems are preferred for large electricity consumers (Ch. 3).

- Know why AC is preferred to DC for long distance power transmission (Ch. 2).

- Be able to design simple inverting, non-inverting amplifiers, integrators, and differentiators using operational amplifiers (Ch. 4).

- Know the characteristics of various electric motors, and when each type is used (Ch. 3).

- Know the characteristics of various transducers and when each type is used (Ch. 7).

- Know the characteristics of a diode and how it is used (Ch. 4).

- Understand computer architecture basics, and the functions of the busses, memory types and computer communication basics (Ch. 5).

- Know what a word, byte, nibble, and bit are (Ch. 5).

- Be able to determine the outputs on a spreadsheet, given the inputs and formula(s) (Ch. 6).

- Know file and record basics (Ch. 6).

- Understand and be able to convert numbers between decimal, binary, octal, and hexadecimal number systems (Ch. 6).

- Be able to follow and determine the result of a computer program flow diagram (Ch. 6).

- Understand the differences, advantages and disadvantages of low level programs, assembly language programs, high level programs, and structured programming (Ch. 6).

- Understand the differences between accuracy, precision, and repeatability (Ch. 7).

FE exam candidates should be aware of the following facts about the exam.

- Circuit analysis problems on the exam only use ideal components (Ch. 1).

- The exam does not include problems on Fourier analysis of waveforms (Ch. 2).

- Exam problems involving complex numbers use the variable j preferred by engineers, rather than the i preferred by mathematicians (Ch. 2).

- Examinees do not need calculus to solve electrical resonance problems on the FE exam (Ch. 2).

- The exam does not require examinees to know the internal details of an op amp beyond the basic schematic diagram (Ch. 4).

- Coding and control systems are not on the current exam, but they could be in the future, and they are an important foundation for computer classes (Ch. 6).

- Examinees do not need calculus to solve maximum power transfer problems on the FE exam (Ch. 1, Ch. 2).

- The NCEES list of topics for the afternoon general portion of the FE exam, effective October 2005, does not include three-phase electricity or electronics.

HOW INSTRUCTORS CAN USE THIS BOOK

FE course instructors are invited to visit PPI's web page (**www.ppi2pass.com/instructorinfo**) for resources for instructors. The instructor's corner contains information on obtaining instructor-only support material, including course outlines, lecture notes, and problem sets, as well as helpful information on course design and exam statistics.

If you are teaching an electrical review course for the FE examination without the benefit of recent, firsthand exam experience, you can use this book as a guide to preparing your lectures. You should spend most of your lecture time discussing the subjects in each chapter.

In solving problems in your lecture, everything you do should be based on the NCEES Handbook. The students can't use your notes in the exam, so train them to use what references they are allowed to use.

It is true that the exam draws upon a body of knowledge that has significantly more breadth than the NCEES Handbook, but few problems will appear that require formulas not present in the handbook. In its attempt to make the FE exam secure, NCEES has been forced to limit the scope of the exam to what is in its handbook.

There are many ways to organize an FE exam review course depending on available time, budget, and audience. However, all good courses have the same result: the students end up storming through the examination.

References

Bogart, Theodore F. Jr. *Electric Circuits*. New York: Macmillan Publishing.

Carr, Joseph J. *Elements of Electronic Instrumentation and Measurement*, 3rd ed. Upper Saddle River, NJ: Prentice Hall.

Cooper, William David. *Electronic Instrumentation and Measurement Techniques*, 2nd ed. Upper Saddle River, NJ: Prentice Hall.

Dorsey, John. *Continuous and Discrete Control Systems*. New York: McGraw-Hill.

Roadstrum, William H., and Dan H. Wolaver. *Electrical Engineering for All Engineers*, 2nd ed. Hoboken, NJ: John Wiley & Sons.

Floyd, Thomas L. *Digital Fundamentals*, 7th ed. Upper Saddle River, NJ: Prentice Hall.

Floyd, Thomas L. *Electronic Devices*, 7th ed. Upper Saddle River, NJ: Prentice Hall.

Johnson, Curtis D. *Process Control Instrumentation Technology*. New York: John Wiley & Sons.

Lindeburg, Michael R. *Engineer-In-Training Reference Manual*. Belmont, CA: Professional Publications, Inc.

Lindeburg, Michael R., Robert R. Angus, John E. Haijar, and Abdulrahman Yassine Haijar. *Electrical Discipline-Specific Review for the FE/EIT Exam*. Belmont, CA: Professional Publications, Inc.

Luppold, David S. *Precision DC Measurements and Standards*. Boston: Addison-Wesley.

Omega Engineering. *Omega Complete Temperature Measurement Handbook and Encyclopedia*, Vol. 27. Stamford, CT: Omega Engineering.

Rizzani, Giorgio. *Principles and Applications of Electrical Engineering*. Boston: Irwin.

Robbins, Allan H., and Wilhelm C. Miller. *Circuit Analysis: Theory and Practice*. Clifton, NY: Thompson Delmar Learning.

Stanley, William D. *Operational Amplifiers with Linear Integrated Circuits*. New York: Macmillan.

Stefani, Raymond, et al. *Design of Feedback Control Systems*. 4th ed. Oxford, UK: Oxford University Press.

Thompson, Lawrence M. *Basic Electrical Measurements and Calibration*. Research Triangle Park, NC: Instrument Society of America.

1 Direct-Current Circuits

Nomenclature

A	area	m^2
A	area: cross section of wire	CM
B	magnetic flux density	T
C	capacitance	F
d	distance, spacing	m
d	diameter	mils
E	electric field strength	V/m
emf	electromotive force	V
F	force	N
G	conductance	S
H	magnetic field intensity	Wb
I	DC current	A
l	length	m
L	inductance	H
N	turns of a coil	–
P	power	W
Q	charge	C
r	distance from wire	m
R	resistance	Ω
t	time	s
V	DC voltage	V
W	work or energy	J or kW·h

Symbols

ε	permittivity	F/m
μ	magnetic permeability	Wb/A·m
ρ	resistivity	Ω·m
α	temperature coefficient	1/°C
θ	magnetic field strength	Wb
τ	time constant	s
Φ	electrical flux	$\dfrac{1}{N}$
Φ	magnetic flux	Wb

Subscripts

N   Norton
Th Thevenin
oc open circuit
sc short circuit
eq equivalent

DIRECT CURRENT

An electrical circuit has a source of electrons and a circuit for the electrons to travel. Basic units of electrical measurement are volts (V), amperes (A), and ohms (Ω). Volts quantify voltage, the force pushing or attracting electrons in the circuit. Amperes quantify current, the rate at which charge (electrons) travels through the circuit. Ohms quantify the circuit's resistance to the flow of electrons. In a *direct-current circuit* (DC circuit), the direction of the current does not change because the voltage does not change polarity.

RESISTANCE

All materials have a resistance, R, to current (electron flow). The amount of resistance is dependent on the type of material and its temperature. Materials with low resistance, such as metals, are categorized as conductors. Materials with high resistance, such as plastics and ceramics, are categorized as insulators and do not permit significant current to travel through them. Materials that have resistances between conductors and insulators are called semiconductors.

Temperature affects a material's resistance. In general, the resistance of conductors increases with an increase in temperature, and the resistance of semiconductors and insulators decreases with an increase in temperature.

Some materials lose their resistance at extremely low temperatures and become superconductors. The FE exam and virtually all electrical engineering and electrical engineering technology courses do not consider superconductors.

In addition to the resistance inherent in the metal wire of a circuit, resistance is often purposely inserted into a circuit to control current using a component called a resistor. Commonly available resistors vary in value from 10 Ω to 22 000 000 Ω. This huge range lends itself to the use of exponents. The electrical industry usually uses units such as kilohms and megohms rather than scientific notation. These units are called engineering units. One kilohm (kΩ) is 1000 Ω and one megohm (MΩ) is 1 000 000 Ω. A milliohm (mΩ) is 0.001 Ω, but this unit is rarely used except for short wires.

RESISTIVITY

Identically shaped resistors of different materials will have different resistances. Resistivity, ρ, is a measure of the electrical resistance of a specific material. Table 1.1 provides resistivity values for some materials. FE exam problems will give resistivity values, so memorizing these is not necessary.

Table 1.1 Resistivities

material	resistivity at 27°C ($\Omega \cdot$m)
silver	1.617×10^{-8}
copper	1.712×10^{-8}
gold	2.255×10^{-8}
aluminum	2.709×10^{-8}
iron	9.87×10^{-8}
nickel	7.8×10^{-8}
lead	21.1×10^{-8}
mercury	96.1×10^{-8}
nichrome	99.72×10^{-8}
carbon	3500×10^{-8}
glass	$1.0 \times 10^{10} - 1.0 \times 10^{14}$
mica	$1.0 \times 10^{11} - 1.0 \times 10^{-15}$

Resistance in a resistor is controlled by the resistivity of its material, its cross sectional area, A, and its length, l. The resistance equation, Eq. 1.1, shows the relationship between resistivity and the other parameters. Resistivity has units of ohm-meters ($\Omega \cdot$m). The meter unit accounts for the cross sectional area and length, and the ohm is the unit of resistance.

$$R = \frac{\rho l}{A} \qquad 1.1$$

Example 1.1

Find the resistance of the nickel bar in the illustration.

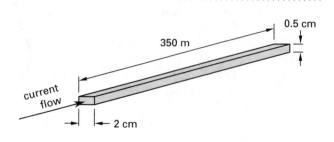

Solution

Use Eq. 1.1 to find the bar's resistance. Obtain the resistivity for nickel from Table 1.1.

$$R = \frac{\rho l}{A}$$

$$= \frac{\left(7.8 \times 10^{-8} \ \Omega \cdot \text{m}\right) \left(350 \ \text{m}\right)}{\left(2 \ \text{cm}\right) \left(0.5 \ \text{cm}\right) \left(100 \ \frac{\text{cm}}{\text{m}}\right)^2}$$

$$= 0.273 \ \Omega$$

Example 1.2

A nichrome wire needs a resistance of 120 Ω. Nichrome is a high-resistance nickel and chromium alloy used for heating elements. Nichrome has a significantly higher resistance than either pure nickel or chromium. The wire has a diameter of 1.0 mm. What is the required wire length?

Solution

Use Eq. 1.1 and solve for length.

$$l = \frac{RA}{\rho} = \frac{R\pi \left(\frac{d}{2}\right)^2}{\rho}$$

$$= \frac{(120 \ \Omega)\pi \left(\frac{1.0 \ \text{mm}}{2}\right)^2 \left(1000 \ \frac{\text{mm}}{\text{m}}\right)^2}{99.72 \times 10^{-8} \ \Omega \cdot \text{m}}$$

$$= 94.5 \ \text{m}$$

SERIES RESISTANCE

The total resistance of resistors in series is the sum of their individual resistances.

$$R_{\text{total}} = R_1 + R_2 + \ldots + R_n \qquad 1.2$$

In Fig. 1.1, the two resistors are identical and in series, so their effective length is twice as long as a single resistor. Therefore the total resistance is twice that of a single resistor.

$$R = \frac{\rho l}{A} + \frac{\rho l}{A}$$

$$= \frac{2\rho l}{A} \qquad 1.3$$

The effective length is twice as long. Therefore the total resistance is twice that of a single resistor. The total series resistance is the sum of the individual resistors.

Example 1.3

If three resistors of 12 Ω, 25 Ω, and 40 Ω are connected in series, what is the total resistance?

Solution

$$R_{\text{total}} = R_1 + R_2 + R_3$$

$$= 12 \ \Omega + 25 \ \Omega + 40 \ \Omega$$

$$= 77 \ \Omega$$

Figure 1.1 Series Resistance

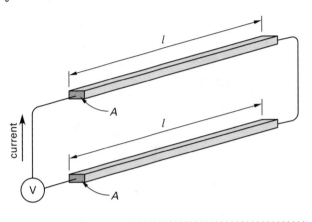

PARALLEL RESISTANCE

A simple parallel circuit has an active source generating the same voltage across two or more parallel circuit legs of the circuit. The total resistance of resistors arranged on these parallel wires, called parallel resistance, is always less than the smallest resistor's value.

Figure 1.2 Simple Parallel Circuit

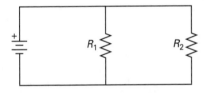

The general equation for parallel resistance is

$$R_{\text{total}} = \frac{1}{\dfrac{1}{R_1} + \dfrac{1}{R_2} + \dfrac{1}{R_3} + \ldots + \dfrac{1}{R_n}} \qquad 1.4$$

The equation for total resistance of two parallel resistors can be derived from the general equation for parallel resistors.

$$R_{\text{total}} = \left(\frac{1}{\dfrac{1}{R_1} + \dfrac{1}{R_2}}\right) \left(\frac{R_1 R_2}{R_1 R_2}\right)$$

$$= \frac{R_1 R_2}{\dfrac{R_1 R_2}{R_1} + \dfrac{R_1 R_2}{R_2}}$$

$$= \frac{R_1 R_2}{R_1 + R_2} \qquad 1.5$$

Figure 1.3 Parallel Resistance

Figure 1.3 Parallel Resistance

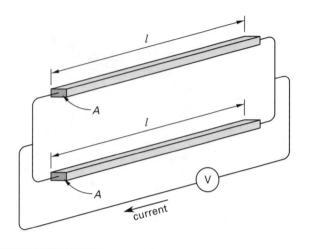

Figure 1.3 shows two identical resistors in parallel. The effective area is twice a single resistor. Therefore the total resistance is half that of a single resistor.

$$R = \frac{\rho l}{2A} \qquad 1.6$$

Example 1.4

What is the total resistance of two 100 Ω resistors in parallel?

Solution

Use Eq. 1.4.

$$R_{\text{total}} = \frac{R_1 R_2}{R_1 + R_2} = \frac{(100 \ \Omega)(100 \ \Omega)}{100 \ \Omega + 100 \ \Omega}$$

$$= 50 \ \Omega$$

Example 1.5

Using the general parallel resistor equation, find the total resistance of two parallel 100 Ω resistors.

Solution

Use Eq. 1.4.

$$R_{\text{total}} = \frac{1}{\dfrac{1}{R_1} + \dfrac{1}{R_2}} = \frac{1}{\dfrac{1}{100 \ \Omega + 100 \ \Omega}}$$

$$= 50 \ \Omega$$

Example 1.6

What is the total resistance of a 12 Ω resistor, a 10 Ω resistor, and a 20 Ω resistor, each on a parallel leg of a circuit?

Solution

Use Eq. 1.4.

$$R_{\text{total}} = \frac{1}{\dfrac{1}{R_1} + \dfrac{1}{R_2} + \dfrac{1}{R_3}}$$

$$= \frac{1}{\dfrac{1}{12 \ \Omega} + \dfrac{1}{10 \ \Omega} + \dfrac{1}{20 \ \Omega}}$$

$$= 4.29 \ \Omega$$

Example 1.7

What is the total resistance of a 100 Ω resistor, 220 Ω resistor, 330 Ω resistor, and 390 Ω resistor in a parallel combination?

Solution

Use Eq. 1.4.

$$R_{\text{total}} = \frac{1}{\dfrac{1}{R_1} + \dfrac{1}{R_2} + \dfrac{1}{R_3} + \dfrac{1}{R_4}}$$

$$= \frac{1}{\dfrac{1}{100 \ \Omega} + \dfrac{1}{220 \ \Omega} + \dfrac{1}{330 \ \Omega} + \dfrac{1}{390 \ \Omega}}$$

$$= 49.65 \ \Omega$$

WIRE UNITS AND GAUGE

The power industry in the United States still uses English units for electrical wire sizes, rather than SI units. It uses circular mils (CM) for the area in the resistance equation. Circular mils are used for circular wires only. A mil is 0.001 in. (Despite the temptation, a mil is never referred to as a milli-inch.) The cross sectional area of a wire in circular mils is equal to the square of the diameter in mils.

$$A = d^2 \quad \text{[circular mils]} \qquad 1.7$$

A common error in calculating resistances is the misuse of π in calculating area: π and resistivity are already factored into circular mils, as shown in the following table of resistivities.

Wire sizes in the United States are made to the American Wire Gage (AWG) standard. Table 1.3 shows the AWG number and the corresponding area in circular mils, diameter in mils, and resistance of copper wire per 1000 ft at 20°C. The resistance values in Table 1.3 are for copper wire, but the dimensions are the same for all wire materials.

Table 1.2 Resistivities for Different Types of Wires

material	resistivity (CM-Ω/ft at 25°C)
silver	9.729
copper	10.30
gold	13.568
aluminum	16.30
tungsten	32.4
nickel	42.84
iron	59.4
lead	127
carbon	2100

Table 1.3 American Wire Gage Sizes

AWG number	area (CM)	approximate diameter (mils)	resistance of copper wire at 20°C (Ω/1000 ft)
0000	211,600	461	0.0491
000	167,810	411	0.0617
00	133,080	364	0.0781
0	105,530	325	0.0982
1	83,694	288	0.1241
2	66,373	258	0.1564
3	52,634	229	0.1971
4	41,742	203	0.2484
5	33,102	183	0.3134
6	26,250	162	0.3952
8	16,509	128	0.6281
10	10,381	103	0.9988
12	6529.0	80.8	1.587
14	4106.8	64.1	2.526
16	2582.9	50.8	4.017
18	1624.3	40.3	6.384
20	1021.5	32.0	10.15
22	642.40	25.3	16.14
24	404.01	20.2	25.67
26	254.10	15.9	40.82
28	159.79	12.7	64.91
30	100.50	10.0	103.1

Example 1.8

Find the resistance of 2500 ft of AWG 14 copper wire at 20°C.

Solution

From Table 1.3, AWG 14 copper wire has a resistance of 2.526 Ω per 1000 ft. The resistance of a 2500 ft length of this wire is

$$R = \rho l = \left(\frac{2.526 \ \Omega}{1000 \ \text{ft}} \right) (2500 \ \text{ft})$$

$$= 6.315 \ \Omega$$

Example 1.9

What is the area in circular mils of a wire with a diameter of 0.015 in?

Solution

Use Eq. 1.7. Convert the given diameter to mils in order to obtain an area in circular mils. A diameter of 0.015 in is equivalent to 15 mils. The area of the wire is

$$A = d^2 = (15 \ \text{mils})^2$$

$$= 225 \ \text{CM}$$

Example 1.10

Find the resistance of 3 mi of AWG 20 aluminum wire.

Solution

Use the resistance equation (Eq. 1.1). From Table 1.3, the diameter of AWG 20 wire is 32.0 mils. The resistivity of aluminum is 16.3 CM-Ω/ft.

$$R = \frac{\rho l}{A_{\text{CM}}}$$

$$= \frac{\left(16.3 \ \dfrac{\text{CM-}\Omega}{\text{ft}} \right) (3 \ \text{mi}) \left(5280 \ \dfrac{\text{ft}}{\text{mi}} \right)}{(32.0 \ \text{mils})^2}$$

$$= 252.14 \ \Omega$$

TEMPERATURE COEFFICIENT OF RESISTANCE

The resistance of any material depends on its temperature. Resistances in tables are usually based on a standard temperature of 20°C (293.15K). Resistance of metals increases linearly with increasing temperature. In contrast, semiconductor resistance decreases non-linearly with increasing temperature.

Linear resistance changes correlate with temperature changes, and can be calculated using Eq. 1.8. The resistance temperature coefficient, α, has units of 1/°C. R_1 is the resistance at the beginning temperature, T_1, and R_2 is the resistance at the final temperature, T_2. ΔT is the difference or change between temperatures T_2 and T_1 (i.e., $T_2 - T_1$).

$$R_2 = R_1(1 + \alpha(T_2 - T_1))$$
$$= R_1(1 + \alpha \Delta T) \qquad 1.8$$

Table 1.4 gives the resistance temperature coefficients of various metals.

The relationship between resistance and temperature means that wire resistance can be used to accurately measure temperature. Metals with low chemical reactivity and high melting points such as platinum are used

Table 1.4 Resistance Temperature Coefficients

metal	resistance temperature coefficient at 20°C (1/°C)
silver	3.80×10^{-3}
copper	3.93×10^{-3}
aluminum	3.91×10^{-3}
tungsten	4.50×10^{-3}
iron	5.50×10^{-3}
lead	4.26×10^{-3}
nichrome	0.44×10^{-3}

to measure temperature in nuclear reactors and other hostile environments. *Thermistors* are semiconductors (usually silicon) used to measure temperatures. Thermistors have resistances that vary greatly with changes in temperature, which make it easy to identify when such a change has occurred, but they are not very accurate. Thermistors are not suitable for high temperatures.

Example 1.11

If the resistance of a copper wire is 150 Ω at 20.0°C, what is the resistance at 200°C? At −50°C?

Solution

Use Eq. 1.8. At 200°C, the wire's resistance is

$$R_2 = R_1 \left(1 + \alpha \left(T_2 - T_1\right)\right)$$

$$= (150 \ \Omega) \left(\begin{array}{l} 1 + \left(3.93 \times 10^{-3} \ \dfrac{1}{°C}\right) \\ \times (200°C - 20°C) \end{array} \right)$$

$$= 256.11 \ \Omega$$

At −50°C, the wire's resistance is

$$R_2 = R_1 (1 + \alpha(T_2 - T_1))$$

$$= (150 \ \Omega) \left(\begin{array}{l} 1 + \left(3.93 \times 10^{-3} \ \dfrac{1}{°C}\right) \\ \times (-50°C - 20°C) \end{array} \right)$$

$$= 108.735 \ \Omega$$

DIRECT CURRENT FLOW AND OHM'S LAW

Which way does direct electrical current flow? Early scientists, Benjamin Franklin among them, believed electricity was a fluid and concluded it flowed from positive to negative. Of course, further research demonstrated that negatively charged electrons are what actually flow, and that they move from negative to positive, but most people still use the old model of current flow. This is called *conventional current flow*. The U.S. military uses *electron current flow* in their training, which is a realistic negative-to-positive model. Figure 1.4 shows the two current flow concepts in a simple DC circuit. Despite the fact that conventional current flow is an inaccurate description of electrical current, either model can be used to calculate voltage, current, and resistance.

Figure 1.4 Conventional Flow vs. Electron Flow

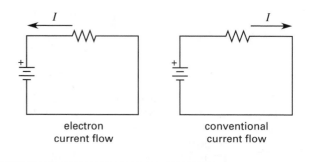

electron
current flow

conventional
current flow

Ohm's law describes the relationship between voltage (in volts), V, current (in amps), I, and resistance (in ohms), R.

$$V = IR \qquad\qquad 1.9$$

The other two versions of Ohm's law are

$$I = \frac{V}{R}$$

$$R = \frac{V}{I}$$

Example 1.12

Given the following voltages and resistances, find the current through the resistor.

(a) 12 V, 30 Ω
(b) 120 V, 40 kΩ
(c) 1.20 kV, 50 kΩ
(d) 1.20 V, 60 kΩ
(e) 150 mV, 30 kΩ

Solution

Use Ohm's law.

(a) $I = \dfrac{V}{R} = \dfrac{12 \ \text{V}}{30 \ \Omega} = 0.4 \ \text{A} \quad (400 \ \text{mA})$

(b) $I = \dfrac{V}{R} = \left(\dfrac{120 \text{ V}}{40 \text{ k}\Omega}\right)\left(\dfrac{1 \text{ k}\Omega}{1000 \text{ }\Omega}\right)$

$\qquad = 0.003 \text{ A} \quad (3.0 \text{ mA})$

(c) $I = \dfrac{V}{R} = \dfrac{1.20 \text{ kV}}{50 \text{ k}\Omega} = 0.024 \text{ A} \quad (24 \text{ mA})$

(d) $I = \dfrac{V}{R} = \left(\dfrac{1.20 \text{ V}}{60 \text{ k}\Omega}\right)\left(\dfrac{1 \text{ k}\Omega}{1000 \text{ k}\Omega}\right)$

$\qquad = 0.000\,02 \text{ A} \quad (20 \text{ }\mu\text{A})$

(e) $I = \dfrac{V}{R} = \left(\dfrac{150 \text{ mV}}{30 \text{ k}\Omega}\right)\left(\dfrac{1 \text{ k}\Omega}{1000 \text{ }\Omega}\right)\left(\dfrac{1 \text{ V}}{1000 \text{ mV}}\right)$

$\qquad = 0.000\,005 \text{ A} \quad (5 \text{ }\mu\text{A})$

Example 1.13

Given the following resistor current and resistance values, find the voltage across the resistor.

(a) 2.2 A, 11.0 Ω
(b) 42 mA, 220 Ω
(c) 0.12 A, 1.3 kΩ
(d) 33 mA, 33 kΩ

Solution
Use Ohm's law.

(a) $\quad V = IR = (2.2 \text{ A})(11.0 \text{ }\Omega) = 24.2 \text{ V}$

(b) $\quad V = IR = (42 \text{ mA})(220 \text{ }\Omega)\left(\dfrac{1 \text{ A}}{1000 \text{ mA}}\right)$

$\qquad = 9.24 \text{ V}$

(c) $\quad V = IR = (0.12 \text{ A})(1.3 \text{ k}\Omega)\left(1000 \text{ }\dfrac{\Omega}{\text{k}\Omega}\right)$

$\qquad = 156 \text{ V}$

(d) $\quad V = IR = (33 \text{ mA})(3.3 \text{ k}\Omega)\left(\dfrac{1 \text{ A}}{1000 \text{ mA}}\right)$

$\qquad \times \left(1000 \text{ }\dfrac{\Omega}{\text{k}\Omega}\right)$

$\qquad = 108.9 \text{ V}$

Example 1.14

Given the following values for voltage across and current through a resistor, find the resistance.
(a) 140 V, 2.4 A
(b) 120 V, 22 mA
(c) 125 V, 2.2 μA
(d) 154 mV, 24 mA

Solution
Use Ohm's law.

(a) $R = \dfrac{V}{I} = \dfrac{140 \text{ V}}{2.4 \text{ A}} = 58.333 \text{ }\Omega$

(b) $R = \dfrac{V}{I} = \left(\dfrac{120 \text{V}}{22 \text{ mA}}\right)\left(\dfrac{1 \text{ A}}{1000 \text{ mA}}\right)$

$\qquad = 5454.54 \text{ }\Omega \quad (5.45 \text{ k}\Omega)$

(c) $R = \dfrac{V}{I} = \left(\dfrac{125 \text{ V}}{2.2 \text{ }\mu\text{A}}\right)\left(1.0 \times 10^6 \text{ }\dfrac{\mu\text{A}}{\text{A}}\right)$

$\qquad = 56\,820\,000 \text{ }\Omega \quad (56.82 \text{ M}\Omega)$

(d) $R = \dfrac{V}{I} = \left(\dfrac{154 \text{ mV}}{24 \text{ mA}}\right)\left(\dfrac{1 \text{ V}}{1000 \text{ mV}}\right)\left(1000 \text{ }\dfrac{\text{mA}}{\text{A}}\right)$

$\qquad = 6.42 \text{ }\Omega$

ELECTRICAL POWER AND ENERGY

Energy is the ability to do work. The units of energy are joules (J). Power is the energy used over time. The units of electrical power are the watt (W), which is equivalent to joules per second (J/s), and the volt-ampere (V·A).

The basic equation for electrical power is used to find the power dissipated (turned to heat) in a resistor or similar device.

$$P = VI \qquad\qquad 1.10$$

Other power equations can be derived from Eq. 1.10 and Ohm's law.

$$P = VI = V\dfrac{V}{R} = \dfrac{V^2}{R} \qquad\qquad 1.11$$

$$P = VI = IRI = I^2 R \qquad\qquad 1.12$$

Mistakes are easy to make when using the power equations that require squaring either the voltage or current. When milliamperes, microamperes, kilovolts, and similar exponential units are squared, exponential errors are likely. The chance of errors can be minimized if the units are reduced to scientific notation (e.g., 1.3×10^3 V or 2.4×10^{-3} A).

Energy (in joules) dissipated in a resistor or other device is found by multiplying power by the time in seconds.

$$\begin{aligned} E_{\text{joules}} &= P_{\text{watts}} t_{\text{seconds}} \\ &= P_{\text{joules/second}} t_{\text{seconds}} \qquad 1.13 \end{aligned}$$

When electric utilities charge their customers for the energy they use, the utilities prefer to use units of kilowatt-hours for energy, rather than joules. The kilowatt-hour is not an SI unit, because the SI unit of time is second, not hour. The basic equation for energy using kilowatt-hours is the same as Eq. 1.11, except for the different units. Energy in kilowatt-hours is equal to power in kilowatts multiplied by the time in hours.

$$E_{\text{kW·hours}} = P_{\text{kW}} t_{\text{hours}} \qquad\qquad 1.14$$

Example 1.15

Find the power dissipated from the following information.

(a) 117 V, 1.1 A
(b) 117 V, 525 mA
(c) 1200 mV, 0.22 A
(d) 115 V, 140 Ω
(e) 0.5 V, 1.2 kΩ
(f) 145 mV, 180 Ω
(g) 2.2 A, 27 Ω
(h) 28 mA, 2.0 kΩ

Solution

Use the different forms of the electrical power equation.

(a) $P = VI = (117 \text{ V})(1.1 \text{ A})$
     $= 128.7 \text{ W}$

(b) $P = VI = (117 \text{ V})(525 \text{ mA})$
     $= 61\,425 \text{ mW} \quad (61.4 \text{ W})$

(c) $P = VI = (1200 \text{ mV})(0.22 \text{ A})$
     $= 264 \text{ mW} \quad (0.264 \text{ W})$

(d) $P = \dfrac{V^2}{R} = \dfrac{(115 \text{ V})^2}{140 \text{ Ω}}$
     $= 94.46 \text{ W}$

(e) $P = \dfrac{V^2}{R} = \dfrac{(0.5 \text{ V})^2}{(1.2 \text{ kΩ})\left(1000 \dfrac{\text{Ω}}{\text{kΩ}}\right)}$

     $= 0.000\,208 \text{ W} \quad (208 \text{ μW})$

(f) $P = \dfrac{V^2}{R} = \left(\dfrac{(145 \text{ mV})^2}{180 \text{ Ω}}\right)\left(\dfrac{1 \text{ V}}{1000 \text{ mV}}\right)^2$

     $= 0.000\,116\,8 \text{ W} \quad (117 \text{ μW})$

(g) $P = I^2 R = (2.2 \text{ A})^2 (27 \text{ Ω})$
     $= 130.68 \text{ W}$

(h) $P = I^2 R = (28 \text{ mA})^2 (2.0 \text{ kΩ})$

     $\times \left(\dfrac{1 \text{ A}}{1000 \text{ mA}}\right)^2 \left(1000 \dfrac{\text{Ω}}{\text{kΩ}}\right)$

     $= 1.568 \text{ W}$

Example 1.16

A 60 W light bulb is on for 1500 s. How much energy has been dissipated in the light bulb?

Solution

$$E = Pt = (60 \text{ W})(1500 \text{ s})$$
$$= \left(60 \ \dfrac{\text{J}}{\text{s}}\right)(1500 \text{ s})$$
$$= 90\,000 \text{ J}$$

Example 1.17

A 15 W light bulb is on for 90 min. How much energy has been dissipated in the light bulb?

Solution

$$E = Pt = (15 \text{ W})(90 \text{ min})$$
$$= \left(15 \ \dfrac{\text{J}}{\text{s}}\right)(90 \text{ min})\left(60 \ \dfrac{\text{s}}{\text{min}}\right)$$
$$= 81\,000 \text{ J}$$

Example 1.18

A 600 W electric space heater is on for 10 h and 30 min. How much energy was used?

Solution

$$E = Pt = (600 \text{ W})(10.5 \text{ h})$$
$$= \left(600 \ \dfrac{\text{J}}{\text{s}}\right)(10.5 \text{ h})\left(60 \ \dfrac{\text{min}}{\text{h}}\right)\left(60 \ \dfrac{\text{s}}{\text{min}}\right)$$
$$= 22\,680\,000 \text{ J} \quad (2.268 \times 10^7 \text{ J})$$

Example 1.19

What is the energy used by a 150 W light bulb operating over 24 h? Express the energy used in joules and kilowatt-hours. How much does the energy cost if the rate is $0.06 per kilowatt-hour?

Solution

The energy in joules is

$$E = Pt = (150 \text{ W})(24 \text{ h})$$
$$= \left(150 \ \dfrac{\text{J}}{\text{s}}\right)(24 \text{ h})\left(60 \ \dfrac{\text{min}}{\text{h}}\right)\left(60 \ \dfrac{\text{s}}{\text{min}}\right)$$
$$= 12\,960\,000 \text{ J} \quad (1.296 \times 10^7 \text{ J})$$

The energy in kilowatt-hours is

$$E = Pt = (150 \text{ W})(24 \text{ h})\left(\dfrac{1 \text{ kW}}{1000 \text{ W}}\right)$$
$$= 3.6 \text{ kW·h}$$

Find the cost of the dissipated energy.

$$(3.6 \text{ kW·h})\left(\dfrac{\$0.06}{1 \text{ kW·h}}\right) = \$0.216$$

Example 1.20

What is the energy used by an electric heater connected to 115 V mains and drawing 11 A over one week? Express the energy dissipated in joules and kilowatt-hours. If the cost per kilowatt-hour is $0.06, what is the total cost of the energy used by the heater?

Solution

Use the basic equation for electrical power.

$$P = VI = (115 \text{ V})(11 \text{ A})$$
$$= 1265 \text{ W} \quad (1.265 \text{ kW})$$

Calculate the energy in joules and kilowatt-hours.

$$E = Pt = (1.265 \text{ kW})(1 \text{ wk})\left(7 \ \frac{\text{d}}{\text{wk}}\right)\left(24 \ \frac{\text{h}}{\text{d}}\right)$$
$$= 212.52 \text{ kW·h}$$

$$E = Pt = (1265 \text{ W})(1 \text{ wk})$$
$$= \left(1265 \ \frac{\text{J}}{\text{s}}\right)(1 \text{ wk})\left(7 \ \frac{\text{d}}{\text{wk}}\right)\left(24 \ \frac{\text{h}}{\text{d}}\right)\left(3600 \ \frac{\text{s}}{\text{h}}\right)$$
$$= 7.65 \times 10^8 \text{ J}$$

Calculate the total cost of the energy.

$$(212.52 \text{ kW·h})\left(\frac{\$0.06}{1 \text{ kW·h}}\right) = \$12.75$$

SERIES CIRCUITS

A series circuit has a source (usually a voltage source) and two or more components in series. For DC circuits, the components are resistors. In AC circuits, the components can be resistors, capacitors, or inductors. Figure 1.5 shows a typical DC series circuit.

The voltage source (also called the voltage rise) is always equal to the voltage across the components (voltage drops). This is *Kirchhoff's voltage law* (KVL). The power given to the circuit by the source must equal the power dissipated by the resistors.

Figure 1.5 Series Circuit

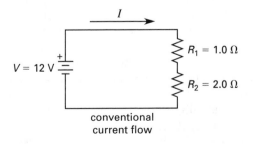

conventional
current flow

In the typical series circuit shown in Fig. 1.5, there is only one path the current can take. The path goes through both resistors R_1 and R_2. The current from the voltage source is the same current passing through each resistor in turn. The current passing through each resistor creates a voltage across that individual resistor, and that voltage can be found through Ohm's law. The subscript i identifies the individual resistance as belonging to a specific resistor. For example, for equations based on the circuit in Fig. 1.5, this subscript would be replaced by the subscripts 1 or 2.

$$V_{R_i} = IR_i$$

The total current must be found to determine these voltages, but first the total resistance has to be calculated using the series resistance equation. Then the current is found by Ohm's law. Finally, the current and the individual resistance values can be plugged into the equation to find the voltage for each resistor.

The *voltage divider rule* is another method of finding the voltage across each resistor in a series circuit. For some circuit problems, input voltage, V_{in}, may be used instead source voltage, V_{source}. They are the same and either notation is correct.

$$V_{R_i} = V_{\text{source}} \frac{R_i}{R_{\text{total}}} \qquad \text{1.15}$$

Example 1.21

For the circuit shown in Fig. 1.5, calculate the individual voltages across each of the circuit's resistors, R_1 and R_2.

Solution

Calculate the total resistance with the equation for series resistance.

$$R_{\text{total}} = R_1 + R_2 = 1.0 \ \Omega + 2.0 \ \Omega$$
$$= 3.0 \ \Omega$$

Using Ohm's law, the current is

$$I = \frac{V}{R_{\text{total}}} = \frac{12 \text{ V}}{3.0 \ \Omega}$$
$$= 4.0 \text{ A}$$

The voltages across each resistor are

$$V_{R_1} = IR_1 = (4.0 \text{ A})(1.0 \ \Omega)$$
$$= 4.0 \text{ V}$$
$$V_{R_2} = IR_2 = (4.0 \text{ A})(2.0 \ \Omega)$$
$$= 8.0 \text{ V}$$

Note that $V_{R_1} + V_{R_2} = 4.0 \text{ V} + 8.0 \text{ V} = 12 \text{ V}$, which is the source voltage, thus satisfying KVL.

Example 1.22

Find the power dissipated in each resistor in the series circuit shown in Fig. 1.5, using the results from the preceding example. Use two methods and compare the total dissipated power to the power supplied by the source.

Solution

The total resistance is

$$R_{\text{total}} = R_1 + R_2 = 1.0 \ \Omega + 2.0 \ \Omega$$
$$= 3.0 \ \Omega$$

From the preceding example, the total current is 4.0 A.

$$I = \frac{V}{R_{\text{total}}} = \frac{12 \ \text{V}}{3.0 \ \Omega}$$
$$= 4.0 \ \text{A}$$

In one approach, the power dissipated in each resistor can be calculated with a version of the power equation.

$$P_{R_1} = I^2 R_1 = (4.0 \ \text{A})^2 (1.0 \ \Omega)$$
$$= 16 \ \text{W}$$
$$P_{R_2} = I^2 R_1 = (4.0 \ \text{A})^2 (2.0 \ \Omega)$$
$$= 32 \ \text{W}$$

For another approach, first calculate the voltages across each resistor.

$$V_{R_1} = IR_1 = (4.0 \ \text{A})^2 (1.0 \ \Omega)$$
$$= 4.0 \ \text{V}$$
$$V_{R_2} = IR_2 = (4.0 \ \text{A})^2 (2.0 \ \Omega)$$
$$= 8.0 \ \text{V}$$

The power dissipated in each resistor can then be calculated with the following power equation.

$$P_{R_1} = \frac{V_{R_1}^2}{R_1} = \frac{(4.0 \ \text{V})^2}{1 \ \Omega}$$
$$= 16 \ \text{W}$$
$$P_{R_2} = \frac{V_{R_2}^2}{R_2} = \frac{(8 \ \text{V})^2}{2.0 \ \Omega}$$
$$= 32 \ \text{W}$$

Add the power dissipated in each resistor to get the total dissipated power.

$$P_{\text{total}} = P_{R_1} + P_{R_2} = 16 \ \text{W} + 32 \ \text{W}$$
$$= 48 \ \text{W}$$

The power supplied by the source is

$$P_{\text{source}} = V_{\text{source}} I = (12 \ \text{V})(4.0 \ \text{A})$$
$$= 48 \ \text{W}$$

Compare this result to the total dissipated power. The power dissipated by the resistors is the same as that supplied by the source: 48 W.

CONDUCTANCE

Conductance is the reciprocal of resistance. It is represented by the variable G and has units of siemens (S).[1]

$$G = \frac{1}{R} \qquad \qquad 1.16$$

The reciprocal relationship between conductance and resistance provides an alternative method for solving circuit problems. This alternative method is particularly useful for parallel circuits.

Example 1.23

Find the conductance of a 12 Ω resistor.

Solution

$$G = \frac{1}{R} = \frac{1}{12 \ \Omega}$$
$$= 0.0833 \ \text{S}$$

PARALLEL CIRCUITS

An example of a parallel circuit is shown in Fig. 1.6.

Figure 1.6 Parallel Circuit

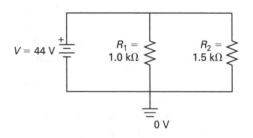

Kirchhoff's current law (KCL) states that the current flowing into a *node (junction)* must equal the current flowing out of that node (junction). A node is a connection or a junction between two or more elements in a circuit. A node that connects two elements is called a simple node. A connection between three or more elements is called a principal node. Elements such as inductors and capacitors might store charge temporarily, but eventually the charge must flow out of the node.

---

[1] Older electrical reference books give the units of conductance as "mho" (i.e., the reverse lettering of ohm, the unit of resistance).

As described in the section on resistance, either Eq. 1.4 or Eq. 1.5 can be used to find total resistance in a parallel circuit. For the circuit in Fig. 1.6, calculate the total current by finding the total parallel resistance. For two parallel resistors, use Eq. 1.5.

$$R_{\text{total}} = \frac{R_1 R_2}{R_1 + R_2} = \frac{(1.0\text{ k}\Omega)(1.5\text{ k}\Omega)}{1.0\text{ k}\Omega + 1.5\text{ k}\Omega}$$
$$= \frac{1.5\text{ k}\Omega^2}{2.5\text{ k}\Omega}$$
$$= 0.6\text{ k}\Omega \quad (600\ \Omega)$$

$$I_{\text{total}} = \frac{V}{R_{\text{total}}} = \frac{44\text{ V}}{0.6\text{ k}\Omega}$$
$$= 73.3\text{ mA}$$

Using Eq. 1.4,

$$R_{\text{total}} = \frac{1}{\dfrac{1}{R_1} + \dfrac{1}{R_2}} = \frac{1}{\dfrac{1}{1.0\text{ k}\Omega} + \dfrac{1}{1.5\text{ k}\Omega}}$$
$$= 0.6\text{ k}\Omega$$

The result for the total resistance is the same using Eq. 1.4 as it is using Eq. 1.5. Thus the total current is the same.

The following equation can be used to find total parallel resistance when all resistors have the same resistance. The number of resistors is represented by n, which for the circuit depicted in Fig. 1.7 would be four.

$$R_{\text{total}} = \frac{R_i}{n} \qquad 1.17$$

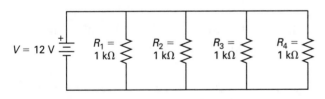

Figure 1.7 Four Equal Parallel Resistors

Example 1.24

Given the circuit shown in Fig. 1.7, what is the total parallel resistance?

Solution

$$R_{\text{total}} = \frac{R_i}{n} = \frac{1000\ \Omega}{4}$$
$$= 250\ \Omega$$

Conductance (the reciprocal of resistance) can also be used to analyze parallel circuits. Some problems lend themselves to easier solutions by using conductance.

$$G = \frac{1}{R} \qquad 1.18$$

The total conductance of a parallel circuit is the sum of the individual conductances.

$$G_{\text{total}} = G_1 + G_2 + ... + G_n \qquad 1.19$$

Using this equation, the total conductance for the circuit in Fig. 1.7 is

$$G_{\text{total}} = \frac{1}{R_1} + \frac{1}{R_2} + \frac{1}{R_3} + \frac{1}{R_4}$$
$$= \frac{1}{1000\ \Omega} + \frac{1}{1000\ \Omega} + \frac{1}{1000\ \Omega} + \frac{1}{1000\ \Omega}$$
$$= \frac{4}{1000\ \Omega}$$
$$= \frac{1}{250\ \Omega}$$
$$= 0.004\text{ S}$$

From this conductance, the total current can be calculated.

$$I_{\text{total}} = VG = V\frac{1}{R}$$
$$= (12\text{ V})(0.004\text{ S})$$
$$= 0.048\text{ A} \quad (48\text{ mA})$$

Example 1.25

What is the total conductance and total resistance of three parallel resistors whose values are 20 Ω, 22 Ω, and 33 Ω?

Solution

Find the conductance values of the individual resistors based on each of their resistances.

$$G_{20\ \Omega} = \frac{1}{20\ \Omega} = 0.050\text{ S}$$
$$G_{22\ \Omega} = \frac{1}{22\ \Omega} = 0.045\text{ S}$$
$$G_{33\ \Omega} = \frac{1}{33\ \Omega} = 0.030\text{ S}$$

Find the total conductance in the circuit.

$$G_{\text{total}} = G_{20\ \Omega} + G_{22\ \Omega} + G_{33\ \Omega}$$
$$= 0.050\text{ S} + 0.045\text{ S} + 0.030\text{ S}$$
$$= 0.125\text{ S}$$

Find the total resistance in the circuit using the total conductance.

$$R_{\text{total}} = \frac{1}{G_{\text{total}}} = \frac{1}{0.125\text{ S}}$$
$$= 8.00\ \Omega$$

Example 1.26

For the parallel circuit in Fig. 1.6, use Ohm's law to find the currents for each resistor and the total current.

Solution

Using Ohm's law, the currents are

$$I_1 = \frac{V}{R_1} = \frac{44 \text{ V}}{1 \text{ k}\Omega}$$
$$= 44 \text{ mA}$$
$$I_2 = \frac{V}{R_2} = \frac{44 \text{ V}}{(1.5 \text{ k}\Omega)\left(1000 \, \frac{\Omega}{\text{k}\Omega}\right)}$$
$$= 0.0293 \text{ A} \quad (29.3 \text{ mA})$$
$$I_{\text{total}} = I_1 + I_2$$
$$= 73.3 \text{ mA}$$

Example 1.27

Using the results from the previous example, find the power dissipated in each resistor (use two methods) and the power supplied by the source (use two methods).

Solution

$$P_{R_1} = I_{R_1}^2 R_1$$
$$= (44 \times 10^{-3} \text{ A})^2 (1 \times 10^3 \, \Omega)$$
$$= 1.936 \text{ W}$$

$$P_{R_2} = I_{R_2}^2 R_2$$
$$= (29.33 \times 10^{-3} \text{ A})^2 (1.5 \times 10^3 \, \Omega)$$
$$= 1.2907 \text{ W}$$

$$P_{\text{total}} = P_{R_1} + P_{R_2}$$
$$= 1.936 \text{ W} + 1.2907 \text{ W}$$
$$= 3.2267 \text{ W} \quad (3.23 \text{ W})$$

$$P_{\text{source}} = V_{\text{source}} I_{\text{total}}$$
$$= (44 \text{ V})(73.33 \text{ mA})$$
$$= 3226.520 \text{ mW} \quad (3.23 \text{ W})$$

Another method of finding the power dissipated and the source power is

$$P_{R_1} = \frac{(V_{R_1})^2}{R_1} = \frac{(44 \text{ V})^2}{(1.0 \text{ k}\Omega)\left(1000 \, \frac{\Omega}{\text{k}\Omega}\right)}$$
$$= 1.936 \text{ W}$$

$$P_{R_2} = \frac{(V_{R_2})^2}{R_2} = \frac{(44 \text{ V})^2}{(1.5 \text{ k}\Omega)\left(1000 \, \frac{\Omega}{\text{k}\Omega}\right)}$$
$$= 1.2907 \text{ W}$$

$$P_{\text{total}} = 1.936 \text{ W} + 1.2907 \text{ W}$$
$$= 3.2267 \text{ W} \quad (3.23 \text{ W})$$

The power is the same using either method.

Example 1.28

For the parallel circuit in Fig. 1.7, first find the total circuit current by adding the current through each of the four resistors. Then find the total current through Ohm's law after determining the total resistance using the parallel resistance equation for identical parallel resistors. Compare the results from these two approaches. Use two different power equations to find both the power dissipated in each resistor and the power supplied by the source.

Solution

Using the first approach to find total current, obtain the current through each resistor.

$$I_{R_i} = \frac{V}{R_i} = \frac{12 \text{ V}}{(1.0 \text{ k}\Omega)\left(1000 \, \frac{\Omega}{\text{k}\Omega}\right)}$$
$$= 0.012 \text{ A} \quad (12 \text{ mA})$$

The total current is the sum of the currents through each resistor.

$$I_{\text{total}} = 12 \text{ mA} + 12 \text{ mA} + 12 \text{ mA} + 12 \text{ mA}$$
$$= 48 \text{ mA}$$

For the second approach to find total current, find the total parallel resistance.

$$R_{\text{total}} = \frac{R_i}{n} = \frac{1.0 \text{ k}\Omega}{4}$$
$$= 0.25 \text{ k}\Omega \quad (250 \, \Omega)$$

Use Ohm's law to find the total current.

$$I_{\text{total}} = \frac{V}{R_{\text{total}}} = \frac{12 \text{ V}}{250 \, \Omega}$$
$$= 0.048 \text{ A} \quad (48 \text{ mA})$$

The currents are the same for both methods.

Find the power dissipated at each resistor and the source power.

$$P_{R_i} = I_{R_i}^2 R_i = (12 \times 10^{-3} \text{ A})^2 (1 \times 10^3 \, \Omega)$$
$$= 0.144 \text{ W}$$

$$P_{\text{total}} = 0.144 \text{ W} + 0.144 \text{ W} + 0.144 \text{ W} + 0.144 \text{ W}$$
$$= 0.576 \text{ mW} \quad (0.58 \text{ W})$$

$$P_{\text{source}} = V_{\text{source}} I_{\text{total}} = (12 \text{ V})(48 \text{ mA})\left(\frac{1 \text{ A}}{1000 \text{ mA}}\right)$$
$$= 0.576 \text{ W} \quad (0.58 \text{ W})$$

Another method of finding the power dissipated and the source power follows.

$$P_{R_i} = \frac{V_{R_i}^2}{R_i} = \frac{(12 \text{ V})^2}{(1.0 \text{ k}\Omega)\left(1000 \, \frac{\Omega}{\text{k}\Omega}\right)}$$
$$= 0.144 \text{ W}$$

$$P_{\text{total}} = 0.144 \text{ W} + 0.144 \text{ W} + 0.144 \text{ W} + 0.144 \text{ W}$$
$$= 0.576 \text{ W} \quad (0.58 \text{ W})$$

$$P_{\text{source}} = \frac{V_{\text{source}}^2}{R_{\text{total}}} = \frac{(12 \text{ V})^2}{(0.25 \text{ k}\Omega)\left(1000 \frac{\Omega}{\text{k}\Omega}\right)}$$

$$= 0.576 \text{ W} \quad (58 \text{ mW})$$

The results are the same using either method.

Example 1.29

Find the total resistance, total current, power in each resistor, total power dissipated, and the total power from the supply for the parallel circuit in the illustration.

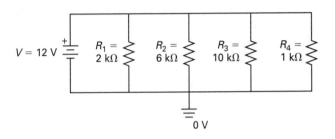

Solution

Find the total parallel resistance.

$$R_{\text{total}} = \frac{1}{\dfrac{1}{R_1} + \dfrac{1}{R_2} + \dfrac{1}{R_3} + \dfrac{1}{R_4}}$$

$$= \frac{1}{\dfrac{1}{2 \text{ k}\Omega} + \dfrac{1}{6 \text{ k}\Omega} + \dfrac{1}{10 \text{ k}\Omega} + \dfrac{1}{1 \text{ k}\Omega}}$$

$$= 0.566 \text{ k}\Omega$$

Find the total current.

$$I_{\text{total}} = \frac{V}{R_{\text{total}}} = \frac{12 \text{ V}}{(0.566 \text{ k}\Omega)\left(1000 \frac{\Omega}{\text{k}\Omega}\right)}$$

$$= 0.0212 \text{ A} \quad (21.2 \text{ mA})$$

Find the current in each resistor, and then sum to find the total current.

$$I_{R_1} = \frac{V}{R_1} = \frac{12 \text{ V}}{2 \text{ k}\Omega}$$

$$= 6 \text{ mA}$$

$$I_{R_2} = \frac{V}{R_2} = \frac{12 \text{ V}}{6 \text{ k}\Omega}$$

$$= 2 \text{ mA}$$

$$I_{R_3} = \frac{V}{R_3} = \frac{12 \text{ V}}{10 \text{ k}\Omega}$$

$$= 1.2 \text{ mA}$$

$$I_{R_4} = \frac{V}{R_4} = \frac{12 \text{ V}}{1 \text{ k}\Omega}$$

$$= 12 \text{ mA}$$

$$I_{\text{total}} = I_{R_1} + I_{R_2} + I_{R_3} + I_{R_4}$$

$$= 6 \text{ mA} + 2 \text{ mA} + 1.2 \text{ mA} + 12 \text{ mA}$$

$$= 21.2 \text{ mA}$$

The total current was found by two methods, and each method agrees.

Find the total power dissipated by the resistor combination. Use two methods to find the total power supplied by the source, then find the power dissipated by each resistor and sum them.

$$P_{\text{total}} = VI_{\text{total}} = (12 \text{ V})(21.2 \text{ mA})$$

$$= 254.4 \text{ mW} \quad (0.25 \text{ W})$$

$$P_{\text{total}} = I_{\text{total}}^2 R_{\text{total}} = (21.2 \text{ mA})^2 (566 \text{ }\Omega)$$

$$= 254.4 \text{ mW} \quad (0.25 \text{ W})$$

$$P_1 = \frac{V^2}{R_1} = \frac{(12 \text{ V})^2}{(2 \text{ k}\Omega)\left(1000 \frac{\Omega}{\text{k}\Omega}\right)}$$

$$= 0.072 \text{ W} \quad (72 \text{ mW})$$

$$P_2 = \frac{V^2}{R_2} = \frac{(12 \text{ V})^2}{(6 \text{ k}\Omega)\left(1000 \frac{\Omega}{\text{k}\Omega}\right)}$$

$$= 0.024 \text{ W} \quad (24 \text{ mW})$$

$$P_3 = \frac{V^2}{R_3} = \frac{(12 \text{ V})^2}{(10 \text{ k}\Omega)\left(1000 \frac{\Omega}{\text{k}\Omega}\right)}$$

$$= 0.014 \text{ W} \quad (14.4 \text{ mW})$$

$$P_4 = \frac{V^2}{R_4} = \frac{(12 \text{ V})^2}{(1 \text{ k}\Omega)\left(1000 \frac{\Omega}{\text{k}\Omega}\right)}$$

$$= 0.144 \text{ W} \quad (144 \text{ mW})$$

$$P_{\text{total}} = 72 \text{ mW} + 24 \text{ mW} + 14.4 \text{ mW} + 144 \text{ mW}$$

$$= 254.4 \text{ mW} \quad (0.25 \text{ W})$$

Power supplied by the source and power dissipated by the resistor combination agree for all methods.

CURRENT DIVIDER RULE

Current entering a node from one direction and exiting into two parallel elements will split, with some current going through each parallel element. Figure 1.8 shows a simple parallel circuit with a node splitting the current in such a manner.

Figure 1.8 Current Divider

Figure 1.8 Current Divider

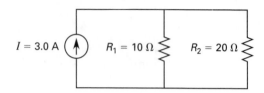

The current divider rule is based on the relationship between the current that enters the node (the input current) and the currents that depart the node through the two downstream elements. The relationship conforms to KCL: currents entering a node equal the currents leaving the node.

$$I_{R_X} = I_{\text{input}} \left(\frac{R_Y}{R_X + R_Y} \right) \qquad 1.20$$

$$I_{R_Y} = I_{\text{input}} \left(\frac{R_X}{R_X + R_Y} \right) \qquad 1.21$$

Applying the current divider rule to a circuit like the one in Fig. 1.8, the value for the R_2 resistor is placed in the numerator to find the current through the R_1 resistor, and the value for the R_1 resistor is placed in the numerator to find the current through the R_2 resistor. This location of resistance values is the opposite of their location in the voltage divider rule equation. The resistance values occupy these spots in the current divider equation because more current goes through the element with the smaller resistance, and less current goes through the element with the larger resistance. Many students confuse this in the exam and make mistakes. A quick way to remember the appropriate arrangement of the current divider equation is to use it to explain a parallel circuit with two legs: one a short circuit (0 Ω resistance) and the other an open circuit (infinite resistance). See Ex. 1.31.

Example 1.30

Find the current through each resistor in the 3.0 A circuit in Fig. 1.8.

Solution
Use the current divider rule to find the current through each resistor.

$$I_{R_1} = I_{\text{input}} \left(\frac{R_2}{R_1 + R_2} \right)$$
$$= (3 \text{ A}) \left(\frac{20 \ \Omega}{10 \ \Omega + 20 \ \Omega} \right)$$
$$= (3 \text{ A}) \left(\frac{20 \ \Omega}{30 \ \Omega} \right)$$
$$= 2 \text{ A}$$

$$I_{R_2} = I_{\text{input}} \left(\frac{R_1}{R_1 + R_2} \right)$$
$$= (3 \text{ A}) \left(\frac{10 \ \Omega}{10 \ \Omega + 20 \ \Omega} \right)$$
$$= (3 \text{ A}) \left(\frac{10 \ \Omega}{30 \ \Omega} \right)$$
$$= 1 \text{ A}$$
$$I_{\text{total}} = I_{R_1} + I_{R_2}$$
$$= 2 \text{ A} + 1 \text{ A}$$
$$= 3 \text{ A}$$

Example 1.31

Find the current through the resistor and the short circuit in the illustration.

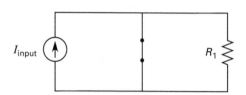

Solution
Use the current divider equation. All the current will go through the short circuit, and none through the resistor.

$$I_{R_1} = I_{\text{input}} \left(\frac{R_{\text{sc}}}{R_1 + R_{\text{sc}}} \right)$$
$$= I_{\text{input}} \left(\frac{0.0 \ \Omega}{R_1 + 0.0 \ \Omega} \right)$$
$$= 0.0 \text{ A}$$

Note the resistor current was found by putting the short circuit resistance into the numerator.

$$I_{R_{\text{sc}}} = I_{\text{input}} \left(\frac{R_1}{R_1 + R_{\text{sc}}} \right)$$
$$= I_{\text{input}} \left(\frac{R_1}{R_1 + 0.0 \ \Omega} \right)$$
$$= I_{\text{input}} \frac{R_1}{R_1}$$
$$= I_{\text{input}}$$

Note the short circuit current was found by putting the resistor's value into the numerator. A common error is to plug the wrong resistance into the numerator of the current divider equation.

SERIES-PARALLEL CIRCUITS

Some circuits have parts that are in series and parts that are in parallel. These circuits are called *series-parallel circuits*. The solution technique is to mathematically convert the circuits to either a completely series equivalent or a completely parallel equivalent. Complicated circuits can be simplified in this manner using equations for series and parallel circuits, including combining series resistances, combining parallel resistances, and combining voltage and current sources (the next section discusses multiple sources in more detail).

Example 1.32

Find the current through each resistor in the series-parallel circuit in the illustration.

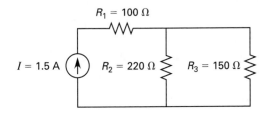

$R_1 = 100\ \Omega$

$I = 1.5\ A$ $R_2 = 220\ \Omega$ $R_3 = 150\ \Omega$

Solution

All the current from I_1 (1.5 A) must pass through R_1; there is no other path. Use the current divider rule to find the currents in R_2 and R_3.

$$I_{R_2} = I_{\text{input}} \left(\frac{R_3}{R_2 + R_3} \right)$$
$$= (1.5\ \text{A}) \left(\frac{150\ \Omega}{370\ \Omega} \right)$$
$$= 0.608\ \text{A}$$
$$I_{R_3} = I_{\text{input}} \left(\frac{R_2}{R_2 + R_3} \right)$$
$$= (1.5\ \text{A}) \left(\frac{220\ \Omega}{220\ \Omega + 150\ \Omega} \right)$$
$$= 0.892\ \text{A}$$

Check the currents through the parallel resistors using KCL: currents entering a node equal the currents exiting the node.

$$I_{\text{total}} = I_{R_2} + I_{R_3}$$
$$= 0.608\ \text{A} + 0.892\ \text{A}$$
$$= 1.5\ \text{A}$$

Example 1.33

Find the total series resistance and the total current for the circuit in the illustration, and find the current through each resistor.

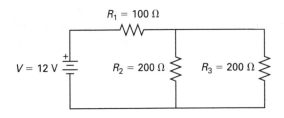

$R_1 = 100\ \Omega$

$V = 12\ V$ $R_2 = 200\ \Omega$ $R_3 = 200\ \Omega$

Solution

Resistors R_2 and R_3 are in parallel. This parallel resistor combination is in series with R_1. The best procedure is to convert the parallel resistor combination to a single equivalent resistor and find the series resistance.

$$R_{\text{parallel}} = \frac{R_2 R_3}{R_2 + R_3} = \frac{(200\ \Omega)(200\ \Omega)}{200\ \Omega + 200\ \Omega}$$
$$= 100\ \Omega$$
$$R_{\text{total}} = R_1 + R_{\text{parallel}} = 100\ \Omega + 100\ \Omega$$
$$= 200\ \Omega$$
$$I_{\text{total}} = \frac{V}{R_{\text{total}}} = \frac{12\ \text{V}}{200\ \Omega}$$
$$= 0.06\ \text{A}\quad (60\ \text{mA})$$

I_{total} is the total current traveling through resistor R_1. I_{total} splits between parallel resistors R_2 and R_3. Since R_2 and R_3 are equal resistance, the currents through the two parallel resistors are equal: 30 mA for each resistor. This can be confirmed by the current divider law.

$$I_{R_2} = I_{\text{total}} \left(\frac{R_3}{R_2 + R_3} \right)$$
$$= (60\ \text{mA}) \left(\frac{100\ \Omega}{200\ \Omega} \right)$$
$$= 30\ \text{mA}$$
$$I_{R_3} = I_{\text{total}} \left(\frac{R_2}{R_2 + R_3} \right)$$
$$= (60\ \text{mA}) \left(\frac{100\ \Omega}{100\ \Omega + 100\ \Omega} \right)$$
$$= 30\ \text{mA}$$

Note the sum of the currents flowing through the two parallel resistors R_2 and R_3 add up to the total circuit current.

Example 1.34

Find the total series resistance, total current, and the current through each resistor in the circuit shown.

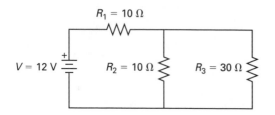

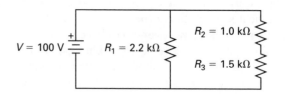

Solution

$$R_{\text{parallel}} = \frac{R_2 R_3}{R_2 + R_3} = \frac{(10 \ \Omega)(30 \ \Omega)}{10 \ \Omega + 30 \ \Omega}$$

$$= 7.5 \ \Omega$$

$$R_{\text{total}} = R_1 + R_{\text{parallel}} = 10 \ \Omega + 7.5 \ \Omega$$
$$= 17.5 \ \Omega$$

$$I_{\text{total}} = \frac{V}{R_{\text{total}}} = \frac{10 \ \text{V}}{17.5 \ \Omega}$$
$$= 0.571429 \ \text{A} \quad (571.4 \ \text{mA})$$

$$I_{R_2} = I_{\text{total}} \frac{R_3}{R_2 + R_3}$$
$$= (571.4 \ \text{mA}) \left(\frac{30 \ \Omega}{10 \ \Omega + 30 \ \Omega} \right)$$
$$= 428.6 \ \text{mA}$$

$$I_{R_3} = I_{\text{total}} \frac{R_2}{R_2 + R_3}$$
$$= (571.4 \ \text{mA}) \left(\frac{10 \ \Omega}{10 \ \Omega + 30 \ \Omega} \right)$$
$$= 142.8 \ \text{mA}$$

Example 1.35

Find the currents in each leg of the circuit shown. Reduce the resistor combination to a single resistor and find the total current. Compare that to the sum of currents in each leg.

Solution

Combine the series resistors R_2 and R_3 into a single resistor, and then find the combined resistance of R_1 in parallel with the combined R_2 and R_3.

$$R_{\text{series}} = R_2 + R_3$$
$$= 1.0 \ \text{k}\Omega + 1.5 \ \text{k}\Omega$$
$$= 2.5 \ \text{k}\Omega$$

$$R_{\text{total}} = \frac{R_1 R_{\text{series}}}{R_1 + R_{\text{series}}}$$
$$= \frac{(2.2 \ \text{k}\Omega)(2.5 \ \text{k}\Omega)}{2.2 \ \text{k}\Omega + 2.5 \ \text{k}\Omega}$$
$$= 1.70 \ \text{k}\Omega$$

$$I_{R_1} = \frac{V}{R_1} = \frac{100 \ \text{V}}{(2.2 \ \text{k}\Omega) \left(1000 \ \frac{\Omega}{\text{k}\Omega} \right)}$$
$$= 0.045 \ \text{A} \quad (45 \ \text{mA})$$

$$I_{\text{series}} = \frac{V}{R_{\text{series}}} = \frac{100 \ \text{V}}{(2.5 \ \text{k}\Omega) \left(1000 \ \frac{\Omega}{\text{k}\Omega} \right)}$$
$$= 0.040 \ \text{A} \quad (40 \ \text{mA})$$

$$I_{\text{total}} = I_{R_1} + I_{\text{series}}$$
$$= 45 \ \text{mA} + 40 \ \text{mA}$$
$$= 85 \ \text{mA}$$

$$I_{\text{total}} = \frac{V}{R_{\text{total}}} = \frac{100 \ \text{V}}{(1.170 \ \text{k}\Omega) \left(1000 \ \frac{\Omega}{\text{k}\Omega} \right)}$$
$$= 0.085 \ \text{A} \quad (85 \ \text{mA})$$

SUPERPOSITION ANALYSIS

The examples shown so far use only one source. A method of analyzing multiple-source circuits is superposition. Voltage sources, current sources, resistors, capacitors and inductors are linear devices. Linear devices allow currents through them and voltages across them to be added or subtracted, depending on polarity. Nonlinear circuits cannot be analyzed using superposition. A semiconductor diode is a non-linear device.

The contribution of each source is analyzed in turn. All sources except one are removed and the circuit is analyzed. The analysis is repeated for each source with all others removed. After the circuit has been analyzed for the contribution of each source, the results are summed together. A voltage source is replaced by a short circuit (0 V, 0 Ω), and a current source is replaced by an open circuit (0 A, ∞ Ω).

Example 1.36

For the circuit in the illustration, find the current in R_1. Use superposition.

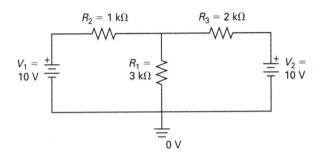

Solution

There are two sources in the illustration, rather than the single sources found in the previous series and parallel circuits. This is not a series or parallel circuit.

By superposition, find the current in R_1. First find the current due to source V_1. Remove V_2 and replace it with a short as shown in the following illustration. R_1 and R_3 are now in parallel and the circuit is now a series-parallel circuit.

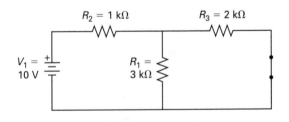

The resistance is

$$R_{V_1 \text{ total}} = R_2 + R_{1,3 \text{ parallel}}$$

$$= 1.0 \text{ k}\Omega + \frac{(2.0 \text{ k}\Omega)(3.0 \text{ k}\Omega)}{2.0 \text{ k}\Omega + 3.0 \text{ k}\Omega}$$

$$= 2.2 \text{ k}\Omega$$

The currents due to source V_1 are

$$I_{V_1 \text{ total}} = \frac{10 \text{ V}}{(2.2 \text{ k}\Omega)\left(1000 \dfrac{\Omega}{\text{k}\Omega}\right)}$$

$$= 0.0045 \text{ A} \quad (4.5 \text{ mA})$$

$$I_{V_1 R_3} = I_{V_1 \text{ total}} \frac{R_1}{R_1 + R_3}$$

$$= (4.5 \text{ mA})\left(\frac{3.0 \text{ k}\Omega}{3.0 \text{ k}\Omega + 2.0 \text{ k}\Omega}\right)$$

$$= 2.7 \text{ mA}$$

$$I_{V_1 R_1} = I_{V_1 \text{ total}} \frac{R_3}{R_1 + R_3}$$

$$= (4.5 \text{ mA})\left(\frac{2.0 \text{ k}\Omega}{3.0 \text{ k}\Omega + 2.0 \text{ k}\Omega}\right)$$

$$= 1.8 \text{ mA}$$

The next step is to find the current in R_1 and then find the current due to source V_2. Remove V_1 from the original circuit and replace it with a short as shown in the following illustration. R_1 and R_3 are now in parallel and the circuit is now a series-parallel circuit.

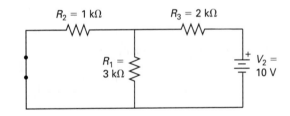

The resistance is

$$R_{V_2 \text{ total}} = R_3 + R_{1,2 \text{ parallel}}$$

$$= 2.0 \text{ k}\Omega + \frac{(1.0 \text{ k}\Omega)(3.0 \text{ k}\Omega)}{1.0 \text{ k}\Omega + 3.0 \text{ k}\Omega}$$

$$= 2.75 \text{ k}\Omega$$

The currents due to source V_2 are

$$I_{V_2 \text{ total}} = \frac{10 \text{ V}}{(2.75 \text{ k}\Omega)\left(1000 \dfrac{\Omega}{\text{k}\Omega}\right)}$$

$$= 0.003636 \text{ A} \quad (3.636 \text{ mA})$$

$$I_{V_2 R_1} = I_{V_2 \text{ total}} \frac{R_2}{R_2 + R_1}$$

$$= (3.636 \text{ mA})\left(\frac{1.0 \text{ k}\Omega}{3.0 \text{ k}\Omega + 1.0 \text{ k}\Omega}\right)$$

$$= 0.909 \text{ mA}$$

The total current in resistor R_1 is

$$I_{R_1 \text{ total}} = I_{V_1 R_1} + I_{V_2 R_1}$$
$$= 1.80 \text{ mA} + 0.909 \text{ mA}$$
$$= 2.709 \text{ mA} \quad (2.7 \text{ mA})$$

Example 1.37

For the circuit illustrated, find the current in R_3 and the voltage at point X. Use superposition.

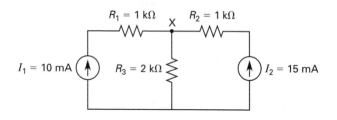

Solution

The superposition technique for current sources is to remove each current source in turn and analyze. Remove current source I_2 and replace it with an open circuit.

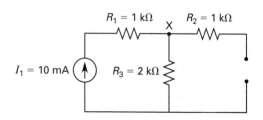

In the preceding illustration, the current in each resistor is

$$I_{1,R_1} = 10 \text{ mA}$$
$$I_{1,R_3} = 10 \text{ mA}$$
$$I_{1,R_2} = 0 \text{ mA}$$

No current flows through R_2.

$$V_{R_1} = (10 \text{ mA})(1.0 \text{ k}\Omega) = 10 \text{ V}$$
$$V_{R_3} = (10 \text{ mA})(2.0 \text{ k}\Omega) = 20 \text{ V}$$

Removing I_1 and replacing it with an open circuit, the resistor currents are

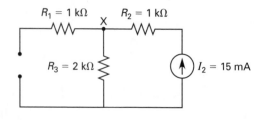

$$I_{2,R_1} = 0 \text{ mA}$$
$$I_{2,R_2} = 15 \text{ mA}$$
$$I_{2,R_3} = 15 \text{ mA}$$
$$V_{R_1} = (15 \text{ mA})(1.0 \text{ k}\Omega)$$
$$= 15 \text{ V}$$
$$V_{R_3} = (15 \text{ mA})(2.0 \text{ k}\Omega)$$
$$= 30 \text{ V}$$

No current flows through R_1. At point X, the voltage is

$$V_X = 20 \text{ V} + 30 \text{ V}$$
$$= 50 \text{ V}$$
$$V_{I_1} = V_X + I_{R_1} R_1$$
$$= 50 \text{ V} + (10 \text{ mA})(1.0 \text{ k}\Omega)\left(1000 \frac{\Omega}{\text{k}\Omega}\right)\left(\frac{1 \text{ A}}{1000 \text{ mA}}\right)$$
$$= 60 \text{ V}$$
$$V_{I_2} = V_X + I_{R_2} R_2$$
$$= 50 \text{ V} + (15 \text{ mA})(1.0 \text{ k}\Omega)\left(1000 \frac{\Omega}{\text{k}\Omega}\right)\left(\frac{1 \text{ A}}{1000 \text{ mA}}\right)$$
$$= 65 \text{ V}$$

The total current in R_3 is

$$I_{R_3} = I_{1,R_3} + I_{2,R_3}$$
$$= 10 \text{ mA} + 15 \text{ mA}$$
$$= 25 \text{ mA}$$

Example 1.38

For the circuit illustrated, use superposition to find the currents in all resistors and the voltage at point X, the junction of the resistors.

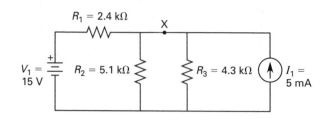

Solution

Remove current source I_1, replace it with an open circuit, and analyze.

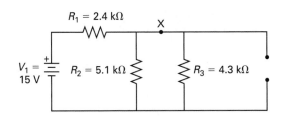

$$R_{\text{total}} = R_1 + R_{2,3 \text{ parallel}}$$

$$= 2.4 \text{ k}\Omega + \frac{(5.1 \text{ k}\Omega)(4.3 \text{ k}\Omega)}{5.1 \text{ k}\Omega + 4.3 \text{ k}\Omega}$$

$$= 4.73 \text{ k}\Omega$$

$$I_{\text{total}} = \frac{15 \text{ V}}{(4.73 \text{ k}\Omega)\left(1000 \frac{\Omega}{\text{k}\Omega}\right)}$$

$$= 0.00317 \text{ A} \quad (3.17 \text{ mA})$$

$$I_{\text{total}} = I_{R_1}$$

By the current divider rule,

$$I_{R_2} = (3.17 \text{ mA})\left(\frac{4.3 \text{ k}\Omega}{4.3 \text{ k}\Omega + 5.1 \text{ k}\Omega}\right)$$

$$= 1.45 \text{ mA}$$

$$I_{R_3} = (3.17 \text{ mA})\left(\frac{5.1 \text{ k}\Omega}{4.3 \text{ k}\Omega + 5.1 \text{ k}\Omega}\right)$$

$$= 1.72 \text{ mA}$$

In the illustration, the source V_1 is replaced by a short.

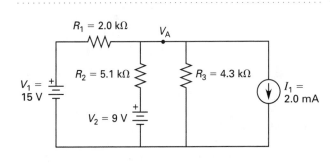

$$R_{\text{total}} = \frac{1}{\dfrac{1}{R_1} + \dfrac{1}{R_2} + \dfrac{1}{R_3}}$$

$$= \frac{1}{\dfrac{1}{2.4 \text{ k}\Omega} + \dfrac{1}{5.1 \text{ k}\Omega} + \dfrac{1}{4.3 \text{ k}\Omega}}$$

$$= 1.18 \text{ k}\Omega$$

$$V_{X,I_1} = (1.18 \text{ k}\Omega)(5 \text{ mA})\left(1000 \frac{\Omega}{\text{k}\Omega}\right)\left(\frac{1 \text{ A}}{1000 \text{ mA}}\right)$$

$$= 5.9 \text{ V}$$

$$I_{R_1} = \frac{5.9 \text{ V}}{(2.4 \text{ k}\Omega)\left(1000 \dfrac{\Omega}{\text{k}\Omega}\right)} = 0.002458 \text{ A} \quad (2.46 \text{ mA})$$

$$I_{R_2} = \frac{5.9 \text{ V}}{(5.1 \text{ k}\Omega)\left(1000 \dfrac{\Omega}{\text{k}\Omega}\right)} = 0.001157 \text{ A} \quad (1.16 \text{ mA})$$

$$I_{R_3} = \frac{5.9 \text{ V}}{(4.3 \text{ k}\Omega)\left(1000 \dfrac{\Omega}{\text{k}\Omega}\right)} = 0.001372 \text{ A} \quad (1.37 \text{ mA})$$

Find the current in each resistor and the voltage at point X from both sources.

$$I_{R_1} = 3.17 \text{ mA} - 2.46 \text{ mA}$$

$$= 0.71 \text{ mA}$$

$$V_{R_1} = (0.71 \text{ mA})(2.4 \text{ k}\Omega)\left(1000 \frac{\Omega}{\text{k}\Omega}\right)\left(\frac{1 \text{ A}}{1000 \text{ mA}}\right)$$

$$= 1.69 \text{ V}$$

$$I_{R_2} = 1.45 \text{ mA} + 1.16 \text{ mA}$$

$$= 2.61 \text{ mA}$$

$$V_{R_2} = (2.61 \text{ mA})(5.1 \text{ k}\Omega)\left(1000 \frac{\Omega}{\text{k}\Omega}\right)\left(\frac{1 \text{ A}}{1000 \text{ mA}}\right)$$

$$= 13.295 \text{ V} \quad (13.30 \text{ V})$$

$$I_{R_3} = 1.72 \text{ mA} + 1.37 \text{ mA}$$

$$= 3.09 \text{ mA} \quad (3.1 \text{ mA})$$

$$V_{R_3} = (3.1 \text{ mA})(4.3 \text{ k}\Omega)\left(1000 \frac{\Omega}{\text{k}\Omega}\right)\left(\frac{1 \text{ A}}{1000 \text{ mA}}\right)$$

$$= 13.296 \text{ V} \quad (13.30 \text{ V})$$

Based on the voltages at R_2 and R_3, the voltage at point X is

$$V_X = 13.30 \text{ V}$$

Example 1.39

Analyze the circuit in the illustration for current and voltage using superposition.

Solution

There are three sources in the problem illustration. Superposition analysis uses one source for each part of the analysis. Two sources will be removed for each portion of the analysis, and the circuit will be analyzed using the one remaining source. V_1 and V_2 will first be replaced by shorts, and the circuit will be analyzed for the contribution of I_1, as shown in the following illustration.

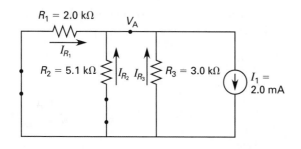

V_1 and V_2 are replaced by short circuits, leaving the resistors in parallel. The parallel equivalent resistance is

$$R_{parallel} = \cfrac{1}{\cfrac{1}{2 \text{ k}\Omega} + \cfrac{1}{5.1 \text{ k}\Omega} + \cfrac{1}{3 \text{ k}\Omega}}$$

$$= 0.9714 \text{ k}\Omega$$

$$V_{total} = (-2 \text{ mA})(0.9714 \text{ k}\Omega)$$

$$= -1.9428 \text{ V}$$

$$I_{R_1} = \cfrac{-1.9428 \text{ V}}{(2.0 \text{ k}\Omega)\left(1000 \cfrac{\Omega}{\text{k}\Omega}\right)}$$

$$= -0.000\,971 \text{ A} \quad (-0.971 \text{ mA})$$

$$I_{R_2} = \cfrac{-1.9428 \text{ V}}{(5.1 \text{ k}\Omega)\left(1000 \cfrac{\Omega}{\text{k}\Omega}\right)}$$

$$= -0.000\,380\,9 \text{ A} \quad (-0.381 \text{ mA})$$

$$I_{R_3} = \cfrac{-1.9428 \text{ V}}{(3.0 \text{ k}\Omega)\left(1000 \cfrac{\Omega}{\text{k}\Omega}\right)}$$

$$= -0.000\,647\,6 \text{ A} \quad (-0.648 \text{ mA})$$

$$I_{R_1} + I_{R_2} + I_{R_4} = -0.971 \text{ mA} - 0.381 \text{ mA} - 0.648 \text{ mA}$$

$$= 2.0 \text{ mA}$$

This equals the current input.

In the next illustration, V_2 has been replaced by a short, and I_1 has been replaced with an open circuit.

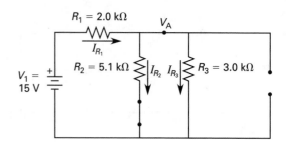

$$R_{total} = 2.0 \text{ k}\Omega + \cfrac{(5.1 \text{ k}\Omega)(3.0 \text{ k}\Omega)}{5.1 \text{ k}\Omega + 3.0 \text{ k}\Omega}$$

$$= 3.889 \text{ k}\Omega$$

$$I_{total} = I_{R_1} = \cfrac{15 \text{ V}}{(3.9 \text{ k}\Omega)\left(1000 \cfrac{\Omega}{\text{k}\Omega}\right)}$$

$$= 0.003846 \text{ A} \quad (3.846 \text{ mA})$$

$$I_{R_2} = (3.846 \text{ mA})\left(\cfrac{3.0 \text{ k}\Omega}{3.0 \text{ k}\Omega + 5.1 \text{ k}\Omega}\right)$$

$$= 1.424 \text{ mA}$$

$$I_{R_3} = (3.846 \text{ mA})\left(\cfrac{5.1 \text{ k}\Omega}{3.0 \text{ k}\Omega + 5.1 \text{ k}\Omega}\right)$$

$$= 2.421 \text{ mA}$$

In the next illustration, V_1 has been replaced with a short, and I_1 has been replaced with an open circuit. The circuit will be analyzed using V_2 voltage source.

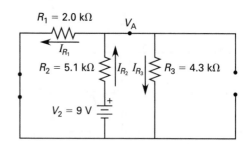

$$R_{total} = 5.1 \text{ k}\Omega + \cfrac{(2.0 \text{ k}\Omega)(3.0 \text{ k}\Omega)}{2.0 \text{ k}\Omega + 3.0 \text{ k}\Omega}$$

$$= 6.3 \text{ k}\Omega$$

$$I_{total} = I_{R_2}$$

$$= \cfrac{9.0 \text{ V}}{(6.3 \text{ k}\Omega)\left(1000 \cfrac{\Omega}{\text{k}\Omega}\right)}$$

$$= 0.001\,429 \text{ A} \quad (1.429 \text{ mA})$$

By the current divider rule,

$$I_{R_1} = I_{\text{total}} \frac{R_3}{R_1 + R_3}$$

$$= (1.429 \text{ mA}) \left(\frac{3.0 \text{ k}\Omega}{2.0 \text{ k}\Omega + 3.0 \text{ k}\Omega} \right)$$

$$= 0.857 \text{ mA}$$

$$I_{R_3} = I_{\text{total}} \frac{R_1}{R_1 + R_3}$$

$$= (1.429 \text{ mA}) \left(\frac{2.0 \text{ k}\Omega}{2.0 \text{ k}\Omega + 3.0 \text{ k}\Omega} \right)$$

$$= 0.572 \text{ mA}$$

The currents have been calculated, but caution must be applied to get the current directions correct. It is helpful to draw arrows on the circuit diagram showing the current direction as each source is analyzed, as shown in the following illustration and other illustrations of this solution.

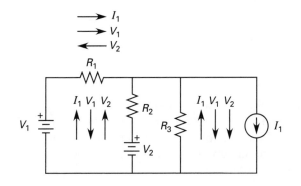

$$\Sigma_{R_1} = I_{I_1} + I_{V_1} + I_{V_2}$$

$$= 0.971 \text{ mA} + 3.846 \text{ mA} - 0.857 \text{ mA}$$

$$= 3.96 \text{ mA} \quad (\text{left to right})$$

$$\Sigma_{R_2} = I_{I_1} + I_{V_1} + I_{V_2}$$

$$= -0.3846 \text{ mA} + 1.429 \text{ mA} - 1.429 \text{ mA}$$

$$= -0.3846 \text{ mA} \quad (\text{top to bottom})$$

$$\Sigma_{R_3} = I_{I_1} + I_{V_1} + I_{V_2}$$

$$= -0.648 \text{ mA} + 2.421 \text{ mA} + 0.572 \text{ mA}$$

$$= 2.345 \text{ mA} \quad (2.3 \text{ mA}) \quad (\text{top to bottom})$$

$$V_{R_1} = (3.96 \text{ mA}) (2.0 \text{ k}\Omega) \left(1000 \frac{\Omega}{\text{k}\Omega} \right) \left(\frac{1 \text{ A}}{1000 \text{ mA}} \right)$$

$$= 7.92 \text{ V}$$

$$V_{R_2} = (-0.3846 \text{ mA}) (5.1 \text{ k}\Omega)$$

$$\times \left(1000 \frac{\Omega}{\text{k}\Omega} \right) \left(\frac{1 \text{ A}}{1000 \text{ mA}} \right)$$

$$= -1.961 \text{V} \quad (-1.96 \text{ V})$$

$$V_{R_3} = V_A = (2.3 \text{ mA}) (3.0 \text{ k}\Omega) \left(1000 \frac{\Omega}{\text{k}\Omega} \right) \left(\frac{1 \text{ A}}{1000 \text{ mA}} \right)$$

$$= 6.9 \text{ V} \quad (7.0 \text{ V})$$

Check if V_A makes sense as the voltage at point A.

$$V_A = V_2 - V_{R_2}$$

$$= 9 \text{ V} - 1.96 \text{ V}$$

$$= 7.04 \text{ V} \quad (7.0 \text{ V})$$

$$V_A = V_1 - V_{R_1}$$

$$= 15 \text{ V} - 7.92 \text{ V}$$

$$= 7.08 \text{ V} \quad (7.0 \text{ V})$$

V_A is the same (7.0 V) relative to each source. It is always wise to do the last step in this solution: a "make sense" check on V_A. Do not blindly trust a calculator. It is only a tool. It is up to the engineer to correctly interpret the calculator's answers.

THEVENIN EQUIVALENT

Thus far in this book, only ideal voltage sources and ideal current sources have been used. An ideal voltage source outputs a constant voltage and has no limit to its output current. A short across an ideal voltage source will draw an infinite current, clearly an impossibility. An ideal current source provides a constant current regardless of its load. If an open circuit is applied across an ideal current source, an infinite voltage will result, also an impossibility.

The voltage at the terminals of real voltage sources, such as batteries, drops as the output current increases. This is modeled by the Thevenin equivalent. The model is an ideal voltage source with a series resistor called the Thevenin equivalent output resistance, R_{Th}, as shown in Fig. 1.9.

Figure 1.9 Thevenin Equivalent

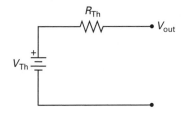

Output current can be measured with an ideal ammeter. An ideal ammeter has 0 Ω internal resistance and acts like a short.

Example 1.40

Reduce the circuit in the illustration to its Thevenin equivalent.

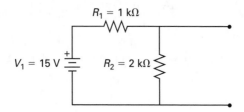

Solution

The problem illustration is the circuit to be analyzed. The no load or open circuit (no current drawn from the circuit) output voltage can be calculated by the voltage divider law to be 10.0 V.

$$V_{\text{Th}} = (15 \text{ V}) \left(\frac{2.0 \text{ k}\Omega}{2.0 \text{ k}\Omega + 1.0 \text{ k}\Omega} \right)$$
$$= 10 \text{ V}$$

In the following illustration, V_1 is replaced by a short and the resistance is measured at the terminals. Resistors R_1 and R_2 are in parallel and the parallel resistance is the Thevenin equivalent resistance.

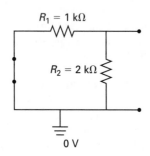

$$R_{\text{Th}} = \frac{R_1 R_2}{R_1 + R_2} = \frac{(1.0 \text{ k}\Omega)(2.0 \text{ k}\Omega)}{1.0 \text{ k}\Omega + 2.0 \text{ k}\Omega}$$
$$= 0.666 \text{ k}\Omega \quad (666 \text{ }\Omega)$$

The following illustration shows the Thevenin equivalent.

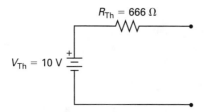

The short-circuit current is found by placing a short circuit across the output.

$$I_{\text{sc}} = \frac{10 \text{ V}}{0.666 \text{ k}\Omega}$$
$$= 15 \text{ mA}$$

This result for current is a finite value.

Example 1.41

Find the Thevenin equivalent of the circuit in the illustration.

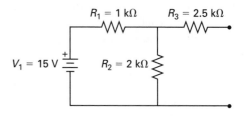

Solution

The difference between the circuit in the last example and the one in this example is the addition of R_3. The Thevenin equivalent voltage has not changed. A common mistake is assuming R_3 changes the Thevenin equivalent voltage.

$$V_{\text{Th}} = (15 \text{ V}) \left(\frac{2.0 \text{ k}\Omega}{2.0 \text{ k}\Omega + 1.0 \text{ k}\Omega} \right)$$
$$= 10 \text{ V}$$

There is no current through R_3 and thus no voltage drop across R_3. The voltage at either end of R_3 is the same. The Thevenin equivalent output voltage is the same as the voltage at the junction of R_1, R_2, and R_3.

However, the Thevenin equivalent resistance and short circuit current do change.

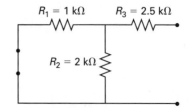

$$R_{\mathrm{Th}} = R_1 \| R_2 + R_3$$
$$= \frac{(1\ \mathrm{k}\Omega)(2\ \mathrm{k}\Omega)}{1\ \mathrm{k}\Omega + 2\ \mathrm{k}\Omega} + 2.5\ \mathrm{k}\Omega$$
$$= 666\ \Omega + 2.5\ \mathrm{k}\Omega$$
$$= 3.166\ \mathrm{k}\Omega$$

$$I_{\mathrm{sc}} = \frac{10\ \mathrm{V}}{(3.1666\ \mathrm{k}\Omega)\left(1000\ \dfrac{\Omega}{\mathrm{k}\Omega}\right)}$$
$$= 0.003\,158\ \mathrm{A} \quad (3.158\ \mathrm{mA})$$

The following illustration is the final Thevenin equivalent.

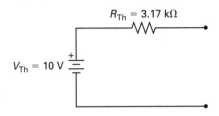

The short-circuit current can be calculated by shorting the circuit output and finding

$$I_{\mathrm{sc}} = I_{R_3}$$

Example 1.42

Find the Thevenin equivalent for the circuit in the illustration.

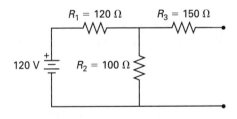

Solution

Find the voltage for the open circuit.

$$V_{\mathrm{oc}} = V_{\mathrm{source}}\left(\frac{R_2}{R_1 + R_2}\right)$$
$$= (120\ \mathrm{V})\left(\frac{100\ \Omega}{100\ \Omega + 120\ \Omega}\right)$$
$$= 54.5\ \mathrm{V}$$

Note that R_3 does not enter into this calculation because no current is drawn through R_3. The open circuit voltage is the same as the voltage across R_2.

To find the Thevenin equivalent resistance we must replace the voltage source with a short circuit, as shown in the following illustration. R_3 is in series with the parallel combination of R_1 and R_2.

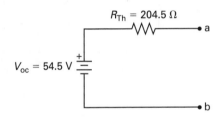

$$R_{\mathrm{Th}} = R_3 + \frac{R_1 R_2}{R_1 + R_2}$$
$$= 150\ \Omega + \frac{(120\ \Omega)(100\ \Omega)}{120\ \Omega + 100\ \Omega}$$
$$= 204.5\ \Omega$$

The following illustration shows the Thevenin equivalent. The short-circuit current is the same as the original circuit.

The short-circuit current is

$$I_{\mathrm{sc}} = \frac{V_{\mathrm{oc}}}{R_{\mathrm{Th}}} = \frac{54.5\ \mathrm{V}}{204.5\ \Omega}$$
$$= 0.267\ \mathrm{A} \quad (267\ \mathrm{mA})$$

The short-circuit current can also be calculated by shorting the circuit output and finding

$$I_{\mathrm{sc}} = I_{R_3}$$

NORTON EQUIVALENT

A Thevenin equivalent is used only with voltage sources. Some analysis methods use current sources and the Norton equivalent is used. The method to find the Norton equivalent is to find the short-circuit current, and then the internal resistance. The internal resistance is found by removing the current source, replacing it with an open circuit, and finding the resistance in the same manner as the Thevenin equivalent resistance. The Norton equivalent resistance is equal to the Thevenin resistance. The Norton equivalent resistance, R_N, is often expressed as conductance because it is often in parallel with another conductance.

Example 1.43

Find the Norton equivalent of the circuit in the illustration. This is the same circuit as in Ex. 1.40.

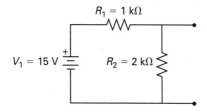

Solution

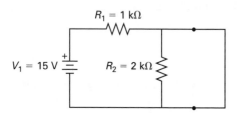

$$I_{sc} = I_N = \frac{15 \text{ V}}{1 \text{ k}\Omega}$$
$$= 15 \text{ mA}$$

The only resistor determining the short-circuit current is R_1. R_2 is shorted out and no current flows through it.

The Norton equivalent resistance is found by removing the voltage source and replacing it with a short, as shown in the following illustration. The Norton equivalent resistance is R_1 in parallel with R_2.

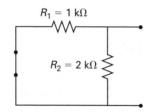

$$R_N = R_{Th} = \frac{R_1 R_2}{R_1 + R_2}$$
$$= \frac{(1 \text{ k}\Omega)(2 \text{ k}\Omega)}{1 \text{ k}\Omega + 2 \text{ k}\Omega}$$
$$= 0.666 \text{ k}\Omega \quad (0.67 \text{ k}\Omega)$$

The following illustration is the final Norton equivalent circuit. The short-circuit current is the current source I_1 output of 15 mA. R_N is shorted out and no current flows through it.

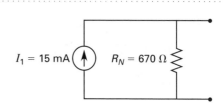

The Norton equivalent open-circuit output voltage is the same as the Thevenin equivalent open-circuit voltage.

$$V_{oc} = V_{Th} = I_1 R_N$$
$$= (15 \text{ mA})(0.67 \text{ k}\Omega)$$
$$= 10 \text{ V}$$

Note the similarities between this solution and Ex. 1.40.

Example 1.44

Find the Norton equivalent of the circuit in the illustration. This is the same circuit as in Ex. 1.41.

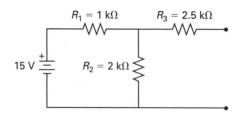

Solution

The Norton equivalent resistance is found by replacing the voltage source with a short and calculating the resistance looking into the terminals as in the following illustration.

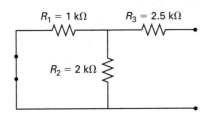

This is exactly the same method as finding the Thevenin equivalent resistance. Indeed the resistances are identical: the Norton equivalent resistance is always the same as the Thevenin equivalent resistance. This is a good check on a circuit analysis.

$$R_N = R_{\text{Th}}$$
$$= R_3 + R_1 \| R_2$$
$$= R_3 + \frac{R_1 R_2}{R_1 + R_2}$$
$$= 2.5 \text{ k}\Omega + \frac{(1.0 \text{ k}\Omega)(2.0 \text{ k}\Omega)}{1.0 \text{ k}\Omega + 2.0 \text{ k}\Omega}$$
$$= 3.166 \text{ k}\Omega$$

I_{sc} is determined by placing a short across the output as in the following illustration. All the current from I_N passes through this short circuit and no current goes through the Norton equivalent resistance.

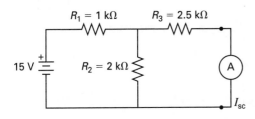

The short-circuit current is

$$R_{\text{total}} = 1 \text{ k}\Omega + \frac{(2 \text{ k}\Omega)(2.5 \text{ k}\Omega)}{2 \text{ k}\Omega + 2.5 \text{ k}\Omega}$$
$$= 2.111 \text{ k}\Omega$$

$$I_{\text{total}} = \frac{V_1}{R_{\text{total}}} = \frac{15 \text{ V}}{(2.111 \text{ k}\Omega)\left(1000 \, \frac{\Omega}{\text{k}\Omega}\right)}$$
$$= 0.007105 \text{ A} \quad (7.105 \text{ mA})$$

By the current divider rule,

$$I_{\text{sc}} = (7.105 \text{ mA})\left(\frac{2.0 \text{ k}\Omega}{2.0 \text{ k}\Omega + 2.5 \text{ k}\Omega}\right)$$
$$= 3.158 \text{ mA}$$

The final Norton equivalent circuit is shown in the following illustration.

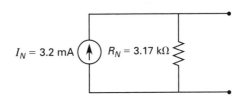

Note the similarities between this solution and Ex. 1.40.

Example 1.45

Find the Norton equivalent of the circuit in the following illustration. This circuit is the same one from Ex. 1.42. Compare the resulting Norton equivalent to the Thevenin equivalent of Ex. 1.42.

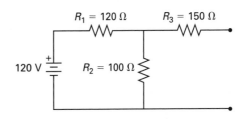

Solution

The short-circuit current is found by measuring the output current through an ammeter, as in the following illustration. An ideal ammeter has 0 Ω internal resistance and acts like a short.

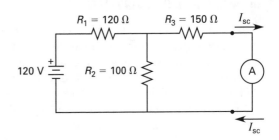

The total resistance seen by the 120 V source is

$$R_{\text{total}} = R_1 + R_2 \| R_3$$
$$= 120 \, \Omega + \frac{(100 \, \Omega)(150 \, \Omega)}{100 \, \Omega + 150 \, \Omega}$$
$$= 120 \, \Omega + 60 \, \Omega$$
$$= 180 \, \Omega$$

The total current from the source is

$$I_{\text{total}} = \frac{120 \text{ V}}{180 \ \Omega}$$
$$= 0.667 \text{ A} \quad (667 \text{ mA})$$

By the current divider rule,

$$I_{\text{sc}} = I_{\text{total}} \frac{R_2}{R_2 + R_3}$$
$$= (667 \text{ mA}) \left(\frac{100 \ \Omega}{100 \ \Omega + 150 \ \Omega} \right)$$
$$= 267 \text{ mA}$$

The R_{internal} is found in the same manner as the Thevenin equivalent resistance. From Ex. 1.42, the Norton equivalent resistance is 204.5 Ω.

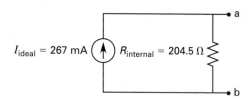

The open-circuit voltage is

$$V_{\text{oc}} = I_{\text{ideal}} R_{\text{internal}}$$
$$= (267 \text{ mA}) (204.5 \ \Omega) \left(\frac{1 \text{ A}}{1000 \text{ mA}} \right)$$
$$= 54.6 \text{ V}$$

This open-circuit voltage is the same for either the Norton or Thevenin equivalent.

NORTON AND THEVENIN EQUIVALENTS

A Norton equivalent and a Thevenin equivalent will look identical to any measurements made at the output terminals. A trick question might ask an engineer to devise a method to distinguish between the two equivalents. The question might read "A black box contains either a Norton equivalent or a Thevenin equivalent: without looking inside the black box, devise a method to determine if there is a Norton or Thevenin equivalent inside." The answer is that because the two equivalents look identical to any external measurements, there is no way to determine which type of equivalent is inside the box.

A Thevenin equivalent is a good model for batteries and power supplies. A Norton equivalent is good for modeling transistors and vacuum tubes. Vacuum tubes are still used in specialized high-power and high-temperature applications.

MAXIMUM POWER TRANSFER

Power transferred from a power source to a load (lights, electric machines, resistive heaters, etc.) depends on the values of the Norton or Thevenin equivalent resistances and the load resistance. Maximum power is transferred from the power source to the load when the Norton or Thevenin equivalent resistance equals the load resistance (i.e., $R_L = R_{\text{Th}}$ and $R_L = R_N$). This demonstrates that the behavior of a Norton equivalent circuit and a Thevenin equivalent circuit are identical.

A common mistake is to think maximum power transfer occurs at maximum current or maximum voltage. Remember, power equals voltage times current. When the voltage is a maximum (load maximum resistance or minimum source resistance) the current is a minimum and power delivered to the load is small. When the current is maximum (load minimum or source resistance maximum) the load voltage is minimum and the power delivered is small. Neither situation gets maximum power transfer.

Figures 1.10 and 1.11 plot power transferred versus load resistance for a source resistance of 204.5 Ω. The Thevenin of Ex. 1.41 is shown in Fig. 1.10. The Norton of Ex. 1.45 is shown in Fig. 1.11. The Norton and Thevenin equivalent resistances are 204.5 Ω. The maximum power transfer occurs when $R_N = R_L$ or $R_{\text{Th}} = R_L$. This demonstrates that the behavior of a Norton equivalent circuit and a Thevenin equivalent circuit is identical.

Figure 1.10 Maximum Power Transfer: Thevenin

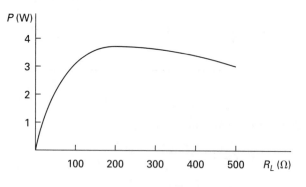

Example 1.46

If a circuit is to drive a load of 51 Ω, what should the Thevenin output resistance be? What resistance value should the Norton output resistance be?

Solution

For maximum power transfer, the Thevenin output resistance and the Norton resistance must equal the load resistance of 51 Ω.

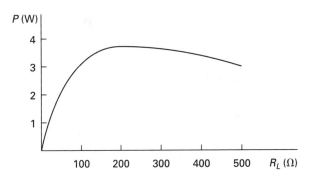

Figure 1.11 Maximum Power Transfer: Norton

$$R_{\text{Th}} = R_N = R_L$$
$$= 51 \ \Omega$$

VOLTMETER LOADING

An analog voltmeter is a microammeter in series with a high value resistor. A microammeter is a very sensitive ammeter. The voltmeter must draw some current to generate a magnetic field that moves the meter pointer. This current draw can affect the voltage reading. If the test circuit output resistance is low, the error will be small. However, if the test circuit output resistance is high, the error will be large. It is desirable to keep the current drawn by the voltmeter as small as possible to prevent changing the circuit behavior. This is done by making the voltmeter internal resistance (R_{meter}) as high as possible. This is not maximum power transfer, but rather the maximum (and most accurate) voltage reading. Modern voltmeters draw little current, but under the wrong circumstances the error could be intolerable.

A voltmeter connected to a circuit acts like a voltage divider as shown in Fig. 1.12.

Figure 1.12 Non-Ideal Voltmeter

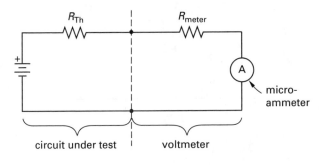

POWER SUPPLY INTERNAL RESISTANCE

Power supplies, whether they are batteries, electronic power supplies, or even a commercial power grid, have an internal resistance. A Thevenin or Norton equivalent can model them and help predict their performance. How is the internal resistance found? It cannot be measured by an ohmmeter. Attempting to do so would severely damage the ohmmeter. The measurement must be done indirectly.

The first step is to measure the open-circuit voltage. An open circuit means no load is attached to the output, and thus no current is drawn from the power supply, as shown in Fig. 1.13.

Figure 1.13 Power Supply Internal Resistance

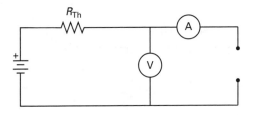

The second step is to put a load on the power supply output and measure the output current and output voltage as shown in Fig. 1.14.

Figure 1.14 Power Supply Internal Resistance

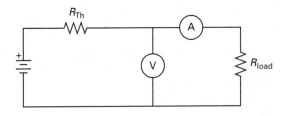

The Thevenin equivalent is found by using the following equation.

$$R_{\text{Th}} = \frac{V_{\text{no load}} - V_{\text{load}}}{I_{\text{load}} - I_{\text{no load}}} \qquad 1.18$$

Example 1.47

A battery open-circuit (no load) voltage is 12.6 V. A load is placed across the battery and draws 1.2 A. The loaded voltage is 12.0 V. What is the Thevenin internal resistance?

Solution

$$R_{\mathrm{Th}} = \frac{V_{\text{no load}} - V_{\text{load}}}{I_{\text{load}} - I_{\text{no load}}}$$

$$= \frac{12.6 \text{ V} - 12.0 \text{ V}}{1.2 \text{ A} - 0.0 \text{ A}}$$

$$= 0.5 \ \Omega$$

MESH (LOOP) ANALYSIS

Mesh, or loop, analysis is useful for multiple source circuits. Its basis is KVL. Thus it works with voltage sources only. The method is best shown by example.

Example 1.48

Find the total current in resistor R_1 in the illustration.

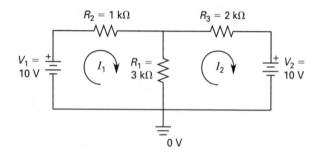

Solution

There are two current loops, I_1 and I_2, disregarding a third loop around the outer edge of the circuit. The current loop direction is chosen arbitrarily, and does not affect the results. Thus two simultaneous algebraic equations are needed. The current loops in resistor R_1 show currents I_1 and I_2 flowing in opposite directions. The total current flowing in R_1 is $I_1 - I_2$. The KVL equations are as follows.

For loop I_1,

$$V_1 = V_{R_2} + V_{R_1}$$
$$= R_2 I_1 + R_1 (I_1 - I_2)$$
$$10 \text{ V} = (1.0 \text{ k}\Omega) I_1 + (3.0 \text{ k}\Omega)(I_1 - I_2)$$
$$= (1.0 \text{ k}\Omega) I_1 + (3.0 \text{ k}\Omega) I_1 - (3.0 \text{ k}\Omega) I_2$$
$$= (4.0 \text{ k}\Omega) I_1 - (3.0 \text{ k}\Omega) I_2$$

For loop I_2,

$$-V_2 = V_{R_3} + V_{R_1}$$
$$= R_3 I_2 + R_1 (I_2 - I_1)$$
$$-10 \text{ V} = (2.0 \text{ k}\Omega) I_2 + (3.0 \text{ k}\Omega)(I_2 - I_1)$$
$$= (2.0 \text{ k}\Omega) I_2 + (3.0 \text{ k}\Omega) I_2 - (3.0 \text{ k}\Omega) I_1$$
$$= (-3.0 \text{ k}\Omega) I_1 + (5.0 \text{ k}\Omega) I_2$$

Combining terms into two simultaneous equations,

$$10 \text{ V} = (4.0 \text{ k}\Omega) I_1 - (3.0 \text{ k}\Omega) I_2$$
$$-10 \text{ V} = (-3.0 \text{ k}\Omega) I_1 + (5.0 \text{ k}\Omega) I_2$$

In matrix form the two simultaneous equations are

$$[R][I] = [V]$$
$$\begin{bmatrix} 4.0 \text{ k}\Omega & -3.0 \text{ k}\Omega \\ -3.0 \text{ k}\Omega & 5.0 \text{ k}\Omega \end{bmatrix} \begin{bmatrix} I_1 \\ I_2 \end{bmatrix} = \begin{bmatrix} V_1 \\ V_2 \end{bmatrix}$$
$$= \begin{bmatrix} 10 \text{ V} \\ -10 \text{ V} \end{bmatrix}$$

Solving the two simultaneous equations,

$$I_1 = -1.818 \text{ mA}$$
$$I_2 = -0.909 \text{ mA}$$

These equations can be solved with Cramer's rule, using computer programs such as MATLAB or Mathcad, or with an adequate calculator.

What is the proper polarity for the sources? The polarity of the source in the equation is the polarity at which the loop arrow leaves the source. The arrow for V_1 leaves V_1's positive terminal. Therefore, V_1 is given a positive sign in the first equation. The arrow for V_2 leaves V_2's negative terminal. Therefore, V_2 is given a negative sign in the second equation.

In the illustration for this problem, there were two sources with their positive symbols at their upper terminals. When these sources were entered into the equations, V_2 had a negative sign.

Example 1.49

In the illustration, find the values for the two loop currents, the current through R_2, and the voltage at the junction of R_1, R_2, and R_3.

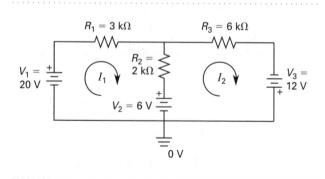

Solution

The two loop equations are

$$20\text{ V} - 6\text{ V} = I_1(3\text{ k}\Omega) + (2\text{ k}\Omega)(I_1 - I_2)$$
$$= I_1(3\text{ k}\Omega) + (2\text{ k}\Omega)I_1 - (2\text{ k}\Omega)I_2$$
$$= I_1(5\text{ k}\Omega) - (2\text{ k}\Omega)I_2$$
$$6\text{ V} + 12\text{ V} = I_2(6\text{ k}\Omega) + (2\text{ k}\Omega)(I_2 - I_1)$$
$$= I_2(6\text{ k}\Omega) + (2\text{ k}\Omega)I_2 - (2\text{ k}\Omega)I_1$$
$$= I_2(8\text{ k}\Omega) - (2\text{ k}\Omega)I_1$$

Note in current loop I_1, the arrow (loop direction) is coming out of V_1's positive terminal and so the 20 V value for V_1 is positive in the first equation. In the same loop, the arrow is coming out of the negative terminal of V_2, and thus the 6 V value of V_2 is negative in the first equation. In the loop I_2, the arrow is coming out of the positive terminals of both V_2 and V_3, and so the source voltages are positive in the equation.

The equations in matrix form are

$$[A][X] = [B] = \begin{bmatrix} a_{1,1} & a_{1,2} \\ a_{2,1} & a_{2,2} \end{bmatrix} \begin{bmatrix} X_1 \\ X_2 \end{bmatrix}$$
$$= \begin{bmatrix} b_1 \\ b_2 \end{bmatrix}$$
$$= [R][I]$$
$$= [V]$$
$$= \begin{bmatrix} 5.0\text{ k}\Omega & -2.0\text{ k}\Omega \\ -2.0\text{ k}\Omega & 8.0\text{ k}\Omega \end{bmatrix} \begin{bmatrix} I_1 \\ I_2 \end{bmatrix}$$
$$= \begin{bmatrix} V_1 \\ V_2 \end{bmatrix}$$
$$= \begin{bmatrix} 14\text{ V} \\ 18\text{ V} \end{bmatrix}$$

Solving the two loop equations,

$$I_1 = 4.111\text{ mA}$$
$$I_2 = 3.27\text{ mA}$$

NODAL ANALYSIS

Nodal analysis uses KCL and current sources, rather than voltage sources. The current entering a node equals the current exiting the node.

Example 1.50

Find voltages V_1 and V_2 in the following illustration.

Solution

$$I_1 = I_{R_2} + I_{R_1}$$
$$I_2 = I_{R_3} + I_{R_1}$$

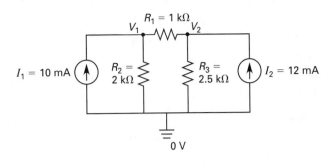

The equation for the first node is

$$10\text{ mA} = \frac{V_1}{R_2} + \frac{V_1 - V_2}{R_1} = \frac{V_1}{2\text{ k}\Omega} + \frac{V_1 - V_2}{1\text{ k}\Omega}$$
$$= \frac{V_1}{2\text{ k}\Omega} + \frac{V_1}{1\text{ k}\Omega} - \frac{V_2}{1\text{ k}\Omega}$$
$$= \left(\frac{0.5 \times 10^{-3}}{1\ \Omega} + \frac{1.0 \times 10^{-3}}{1\ \Omega} \right) V_1$$
$$- \left(\frac{1.0 \times 10^{-3}}{1\ \Omega} \right) V_2$$
$$= \left(\frac{1.5 \times 10^{-3}}{1\ \Omega} \right) V_1 - \left(\frac{1.0 \times 10^{-3}}{1\ \Omega} \right) V_2$$

The equation for the second node is

$$12\text{ mA} = \frac{V_2}{R_3} + \frac{V_2 - V_1}{R_1} = \frac{V_2}{2.5\text{ k}\Omega} + \frac{V_2 - V_1}{1.0\text{ k}\Omega}$$
$$= \left(\frac{0.4 \times 10^{-3}}{1\ \Omega} \right) V_2 + \left(\frac{1.0 \times 10^{-3}}{1\ \Omega} \right)$$
$$\times (V_2 - V_1)$$
$$= \left(\frac{0.4 \times 10^{-3}}{1\ \Omega} \right) V_2 + \left(\frac{1.0 \times 10^{-3}}{1\ \Omega} \right) V_2$$
$$- \left(\frac{1.0 \times 10^{-3}}{1\ \Omega} \right) V_1$$
$$= -\left(\frac{1.0 \times 10^{-3}}{1\ \Omega} \right) V_1 + \left(\frac{1.4 \times 10^{-3}}{1\ \Omega} \right) V_2$$

In matrix form, the equations are

$$[A][X] = [B] = \left[\frac{1}{R} \right][V] = [I]$$
$$= \left[\frac{1}{R} \right] \begin{bmatrix} V_1 \\ V_2 \end{bmatrix}$$
$$= \begin{bmatrix} I_1 \\ I_2 \end{bmatrix}$$
$$= \begin{bmatrix} 1.5 \times 10^{-3} & -1.0 \times 10^{-3} \\ -1.0 \times 10^{-3} & 1.4 \times 10^{-3} \end{bmatrix} \begin{bmatrix} V_1 \\ V_2 \end{bmatrix}$$
$$= \begin{bmatrix} 10 \times 10^3\text{ A} \\ 12 \times 10^3\text{ A} \end{bmatrix}$$

Solving for the two voltages,

$$V_1 = 23.636 \text{ V}$$
$$V_2 = 25.455 \text{ V}$$

These equations can be solved by programs such as MATLAB or Mathcad, with Cramer's rule, or with a suitable calculator.

Example 1.51

Find voltages V_1 and V_2 in the following illustration.

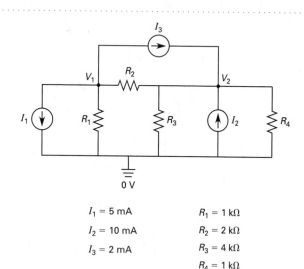

$$I_1 = 5 \text{ mA} \qquad R_1 = 1 \text{ k}\Omega$$
$$I_2 = 10 \text{ mA} \qquad R_2 = 2 \text{ k}\Omega$$
$$I_3 = 2 \text{ mA} \qquad R_3 = 4 \text{ k}\Omega$$
$$\qquad\qquad\qquad R_4 = 1 \text{ k}\Omega$$

Solution

For the first node,

$$-5 \text{ mA} - 2 \text{ mA} = \frac{V_1}{R_1} + \frac{V_1 - V_2}{R_2} = \frac{V_1}{1.0 \text{ k}\Omega} + \frac{V_1 - V_2}{2.0 \text{ k}\Omega}$$

$$-7 \text{ mA} = \left(\frac{1.0 \times 10^{-3} + 0.5 \times 10^{-3}}{1 \, \Omega} \right) V_1$$
$$- \left(\frac{0.5 \times 10^{-3}}{1 \, \Omega} \right) V_2$$
$$= \left(\frac{1.5 \times 10^{-3}}{1 \, \Omega} \right) V_1 - \left(\frac{0.5 \times 10^{-3}}{1 \, \Omega} \right) V_2$$

For the second node,

$$2 \text{ mA} + 10 \text{ mA} = \frac{V_2}{R_3} + \frac{V_2}{R_4} + \frac{V_2 - V_1}{R_2}$$
$$= \frac{V_2}{4.0 \text{ k}\Omega} + \frac{V_2}{1.0 \text{ k}\Omega} + \frac{V_2 - V_1}{2.0 \text{ k}\Omega}$$

$$12 \text{ mA} = - \left(\frac{0.5 \times 10^{-3}}{1 \, \Omega} \right) V_1$$
$$+ \left(\frac{0.25 \times 10^{-3} + 1.0 \times 10^{-3}}{1 \, \Omega} \right) V_2$$

$$= - \left(\frac{0.5 \times 10^{-3}}{1 \, \Omega} \right) V_1 + \left(\frac{1.75 \times 10^{-3}}{1 \, \Omega} \right) V_2$$

Solving,

$$\left[\frac{1}{R} \right] [V] = [I]$$

$$= \begin{bmatrix} \dfrac{1.5 \times 10^{-3}}{1 \, \Omega} & -\dfrac{0.5 \times 10^{-3}}{1 \, \Omega} \\ -\dfrac{0.5 \times 10^{-3}}{1 \, \Omega} & \dfrac{1.75 \times 10^{-3}}{1 \, \Omega} \end{bmatrix} \begin{bmatrix} V_1 \\ V_2 \end{bmatrix}$$

$$= \begin{bmatrix} -7 \times 10^{-3} \text{ A} \\ 12 \times 10^{-3} \text{ A} \end{bmatrix}$$

All values have the 10^{-3} term, so they cancel out. It is not necessary to enter the 10^{-3} into the calculator or computer. Solving for the two voltages,

$$V_1 = -2.632 \text{ V}$$
$$V_2 = 6.105 \text{ V}$$

BRIDGE CIRCUITS

Bridge circuits are used to find the value of an unknown resistance. They are more accurate than most analog ohmmeters. A bridge circuit is two resistive voltage dividers. The voltage at point A is adjusted until V_A and V_B are equal and no current flows between the two voltage dividers. Each voltage divider can be reduced to two Thevenin equivalents for analysis.

Figure 1.15 Bridge Circuit

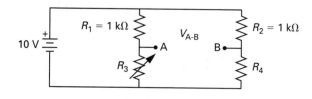

A typical bridge circuit is shown in Fig. 1.15. In the illustration, resistors R_1 and R_2 are at a fixed value and known, and R_3 is a variable resistor with dial readout of its resistance. R_3 is adjusted until V_A equals V_B and current does not flow between points A and B. The value of R_4, an unknown resistance, is found by solving the following equation.

$$R_4 = \frac{R_2 R_3}{R_1} \qquad\qquad 1.19$$

A typical test question is to find the unknown resistance given the other three resistances. Another typical test question is to give resistance values and ask if the bridge is in balance. The following equations are valid only when the bridge is balanced.

$$\frac{R_1}{R_2} = \frac{R_3}{R_4} \qquad\qquad 1.20$$

$$\frac{R_1}{R_3} = \frac{R_2}{R_4} \qquad\qquad 1.21$$

Example 1.52

The bridge circuit in Fig. 1.15 is balanced. $R_2 = 1.0$ kΩ, $R_3 = 2.0$ kΩ, and $R_1 = 1.0$ kΩ. What is the value of resistor R_4?

Solution

$$R_4 = \frac{R_2 R_3}{R_1} = \frac{(1.0 \text{ k}\Omega)(2.0 \text{ k}\Omega)}{1.0 \text{ k}\Omega}$$
$$= 2.0 \text{ k}\Omega$$

This measurement is only valid if $V_{\text{A-B}} = 0$ V. If this condition is met, the bridge is said to be balanced. If the bridge is balanced, no current will flow between points A and B. The balanced condition can be determined by a microammeter. A microammeter is a sensitive ammeter designed to measure small currents. When R_3 is adjusted for no current, the bridge is in balance.

ELECTRICAL FIELDS

An excess of electrons creates a negative charged object. A shortage of electrons creates a positive charged object. An electrical charge creates an electric field. This field affects other charges. Charges of opposite polarity (positive versus negative) attract each other, and charges of the same polarity repel each other. See Fig. 1.16. The force created by the charges is

$$F = k\frac{Q_1 Q_2}{r^2} \qquad 1.22$$

F is the force in newtons, k is a constant 9×10^9 in N·m^2/C^2, Q_1 and Q_2 are the charges in coulombs, and r is the distance between the charges in meters. The only way electric fields can be measured is by measuring this force. Electrical fields cannot exist within electrical conductors.

Figure 1.16 Force between Two Charges

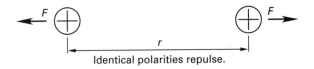

Identical polarities repulse.

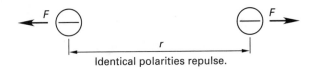

Identical polarities repulse.

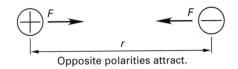

Opposite polarities attract.

The electric field created by a point charge falls off as the inverse square of the distance. This inverse square effect is similar to gravity and light intensity. Most electric field calculations assume a point charge, which means the charge has a very small physical size compared to the separation distance. This greatly simplifies calculations.

Example 1.53

Given two charges of 50 μC and -120 μC separated by 40 mm, find the force created between them.

Solution

$$F = k\frac{Q_1 Q_2}{r^2}$$
$$= \left(9.0 \times 10^9 \ \frac{\text{N} \cdot \text{m}^2}{\text{C}^2}\right)$$
$$\times \left(\frac{(50 \times 10^{-6} \text{ C})(-120 \times 10^{-6} \text{ C})}{(40 \times 10^{-3} \text{ m})^2}\right)$$
$$= -3.375 \times 10^4 \text{ N} \quad [\text{attractive force}]$$

PERMITTIVITY

The permittivity of a material, ε, measures how easily a material will form an electric field, and has units of farads/meter (F/m). It is one of the special constants that are needed to fit electrical units into the same system (System International) as units of time, distance, and charge. Electrical and magnetic units and constants make electric charge, resistance, and magnetic and electric fields all fit together in the SI (metric) system. Time (in seconds) and distance (in meters) are related to the time of the earth's rotation and to the earth's circumference respectively, but not to any magnetic or electrical quantity. The coulomb was defined before electrons were discovered.

The permittivity varies with the material. The absolute permittivity, ε, of a material is the material's relative permittivity, ε_r, (a unitless constant) times the permittivity of a vacuum, ε_0.

$$\varepsilon = \varepsilon_r \varepsilon_0 \qquad 1.23$$

The permittivity of a vacuum is $\varepsilon_0 = 8.84 \times 10^{-12}$ F/m. Table 1.5 lists relative and absolute permittivities.

Permittivity can be used to measure the electric field strength, E, emanating radially from a point charge.

$$E = \frac{Q}{4\pi\varepsilon r^2} \qquad 1.24$$

Table 1.5 Permittivities

material	permittivity, ε (F/m)	relative permittivity, ε_r (unitless dielectric constant)
vacuum	8.84×10^{-12}	1.00
air	8.85×10^{-12}	1.006
Teflon	17.68×10^{-12}	2
paraffined paper	22.10×10^{-12}	2.5
polystyrene	22.98×10^{-12}	2.6
mica	44.20×10^{-12}	5.5
Mylar	26.52×10^{-12}	3
porcelain	53.04×10^{-12}	6
Bakelite	61.88×10^{-12}	7
glass	66.30×10^{-12}	7.5
water (distilled)	707×10^{-12}	80
barium strontium titanite	$66{,}300 \times 10^{-12}$	7500
ceramic (various types)	$265\text{–}66\,300 \times 10^{-12}$	30–7500

Example 1.54

What is the electrical field 10 cm from a point charge of 100 μC in air?

Solution

$$E = \frac{Q}{4\pi\varepsilon r^2}$$

$$= \frac{100 \times 10^{-6}\ \text{C}}{4\pi \left(8.85 \times 10^{-12}\ \dfrac{\text{F}}{\text{m}}\right)(0.1\ \text{m})^2}$$

$$= 89.9 \times 10^6\ \text{C/F·m} \quad (89.9 \times 10^6\ \text{V/m})$$

CAPACITANCE

Capacitance is the capacity to store charge. The charge is stored on a capacitor. A capacitor can be as simple as two parallel conducting plates (usually metal), as shown in Fig. 1.17 with a battery charging the capacitor. The amount of charge stored depends on the capacitance value of the capacitor. Charge, Q, is given in units of coulombs (C). A coulomb is the charge of 6.242×10^{18} electrons. One electron has the following charge.

$$\frac{1\ \text{C}}{6.242 \times 10^{18}\ \text{electrons}} = 1.6022 \times 10^{-19}\ \text{C/electron}$$

Capacitance in farads (F) is equivalent to charge over voltage, which is measured in coulombs per volt (C/V).

$$C = \frac{Q}{V} \qquad 1.25$$

The charge on a capacitor is

$$Q = CV \qquad 1.26$$

One ampere is equal to electron (charge) flow of 1 C/s. A farad is a very large quantity, so microfarad (μF = 10^{-6} F), nanofarad (nF = 10^{-9} F), and picofarad (pF = 10^{-12} F) are normally used.

In a typical capacitor, the plates each have an area in square meters, A, and they are separated from each other by a distance in meters, d, of dielectric (non-conducting material). The permittivity is ε. The capacitance is given by

$$C = \frac{\varepsilon A}{d} = \frac{\varepsilon_r \varepsilon_0 A}{d} \qquad 1.27$$

Figure 1.17 Capacitor

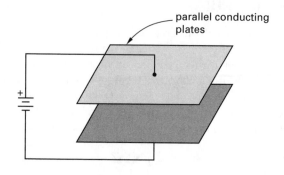

parallel conducting plates

For an air dielectric, the relative permittivity is $\varepsilon_r \approx 1$ and the capacitance, C, is given by

$$C = \frac{\varepsilon A}{d} = \frac{\varepsilon_r \varepsilon_0 A}{d} \approx \frac{\varepsilon_0 A}{d} \qquad 1.28$$

Any dielectric material between the capacitor plates will increase the capacitance. The amount a material increases the relative permittivity is ε_r.

Example 1.55

Find the charge on a 12 μF capacitor charged to 22 V.

Solution

$$Q = CV = (12\ \mu\text{F})\,(22\ \text{V})$$

$$= \left(12 \times 10^{-6}\ \frac{\text{C}}{\text{V}}\right)(22\ \text{V})$$

$$= 264 \times 10^{-6}\ \text{C} \quad (264\ \mu\text{C})$$

Example 1.56

Find the capacitance of an air dielectric capacitor with dimensions of 0.4 m × 0.5 m, and a plate spacing of 1.0 mm.

Solution

From Table 1.5, the relative permittivity of air is 1.006.

$$
\begin{aligned}
C &= \frac{\varepsilon A}{d} = \frac{\varepsilon_r \varepsilon_0 A}{d} \\
&= \frac{(1.006)\left(8.84 \times 10^{-12}\ \frac{\text{F}}{\text{m}}\right)(0.4\ \text{m})(0.5\ \text{m})}{1 \times 10^{-3}\ \text{m}} \\
&= 1.77 \times 10^{-9}\ \text{F}\quad (1.8\ \text{pF})
\end{aligned}
$$

Example 1.57

What would be the area of a 1.0 F capacitor with an air dielectric and a plate spacing of 0.1 mm?

Solution

$$
\begin{aligned}
A &= \frac{Cd}{\varepsilon_0} = \frac{(1.0\ \text{F})\left(10^{-4}\ \text{m}\right)}{8.84 \times 10^{-12}\ \frac{\text{F}}{\text{m}}} \\
&= 11.3 \times 10^{6}\ \text{m}^2
\end{aligned}
$$

The square capacitor would have to be 3.363 km on a side. Not exactly something that could be used in a hand held electronic device. Actually, high-value capacitors are made with high relative-permittivity dielectrics. An additional technique to save volume is to roll the plates into a cylinder, instead of leaving them flat.

Large, high-voltage capacitors are used in powerful lasers, such as those used for hydrogen fusion research at the Los Alamos National Laboratory in Los Alamos, N.M. Power is taken from commercial sources, and charge is stored over a relatively long period of time. The capacitor is discharged into the lasers in a short pulse, creating intense light beams that are sent into a hydrogen target. The needed high voltage requires wide plate spacing to prevent voltage breakdown.

ELECTRIC FIELD IN A CAPACITOR

The charges on a capacitor create an electric field between the plates as shown in Fig. 1.18.

A charge within the capacitor will experience a force due to the electric field. The relationship between this force, F, the electric field intensity, $\overline{E}$, and the charge, Q, is

$$
\overline{E} = \frac{F}{Q} \qquad 1.29
$$

Electrical field intensity units are either newtons/coulomb (N/C) or volts/meter (V/m).

The electric field is assumed to be uniform everywhere within a parallel plate capacitor. Thus the force on a charge does not depend on its location within the capacitor. Actually, the field bows outward at the plate edges, but this is usually ignored for problem simplification.

Electric flux can be visualized by flux lines emanating from positive charges and terminating on negative charges as shown in Fig. 1.18. The flux density, D, is the flux in coulombs, Φ, divided by the area, A.

$$
D = \frac{\Phi}{A} \qquad 1.30
$$

Figure 1.18 Electric Field in a Capacitor

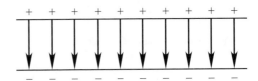

The relationship between the flux density and an electric field with a permittivity, ε, is

$$
D = \varepsilon \overline{E} \qquad 1.31
$$

Example 1.58

What is the electric field intensity if a 50 μC charge within that field experiences a 30 N force?

Solution

$$
\begin{aligned}
\overline{E} &= \frac{F}{Q} = \left(\frac{30\ \text{N}}{50\ \mu\text{C}}\right)\left(1 \times 10^6\ \frac{\mu\text{C}}{\text{C}}\right) \\
&= 0.6 \times 10^6\ \text{N/C}\quad (6.0 \times 10^5\ \text{V/m})
\end{aligned}
$$

Example 1.59

A parallel-plate, air-dielectric capacitor has a total charge of 120 μC with an area of 4 cm^2. Find the flux density and the electric field intensity.

Solution

$$
\begin{aligned}
D &= \frac{\Phi}{A} = \left(\frac{120\ \mu\text{C}}{4 \times 10^{-4}\ \text{m}^2}\right)\left(1 \times 10^6\ \frac{\mu\text{C}}{\text{C}}\right) \\
&= 30 \times 10^{-2}\ \text{C/m}^2\quad \left(0.3\ \text{C/m}^2\right)
\end{aligned}
$$

From Table 1.5, the permittivity of air is 8.85×10^{-12} F/m. Thus,

$$D = \varepsilon \overline{E}$$

$$\overline{E} = \frac{D}{\varepsilon} = \frac{0.3 \, \dfrac{\text{C}}{\text{m}^2}}{8.85 \times 10^{-12} \, \dfrac{\text{F}}{\text{m}}}$$

$$= 3.4 \times 10^{10} \text{ V/m}$$

Work in an Electric Field

Work is done if a charge is moved against the electric field. The work is the force times the distance moved against the field.

$$W = Fd = Q\overline{E}d \qquad \text{1.32}$$

Voltage can be defined as work divided by charge.

$$V = \frac{W}{Q} = \frac{Fd}{Q} \qquad \text{1.33}$$

Dividing both sides by the distance yields

$$\frac{V}{d} = \frac{F}{Q} = \overline{E} \qquad \text{1.34}$$

Example 1.60

A parallel plate capacitor has an electric field of 1200 V/m. A charge of 100 μC is moved 1.0 mm against the field. How much work is done?

Solution

$$W = Fd = Q\overline{E}d$$

$$= \left(100 \times 10^{-6} \text{ C}\right) \left(1200 \, \frac{\text{V}}{\text{m}}\right) \left(1.0 \times 10^{-3} \text{ m}\right)$$

$$= 1.2 \times 10^{-4} \text{ J}$$

VOLTAGE BREAKDOWN

In a charged capacitor, a voltage difference exists between the capacitor plates, creating an electric field between the plates. The electric field is equal to the voltage between the plates (in volts) divided by the distance between the plates (in meters).

$$\overline{E} = \frac{V}{d} \qquad \text{1.35}$$

There is a limit to a capacitor's electric field, and this limit is called the breakdown strength. The breakdown strength is dependent on the dielectric. Table 1.6 shows the breakdown strength of various dielectrics. The breakdown voltages in the table are only approximate. Many factors affect the actual breakdown strength, such as moisture, manufacturing quality, and temperature

Table 1.6 Typical Breakdown Voltages

material	breakdown voltage (V/m)
air	3×10^6
porcelain	7×10^6
oil	15×10^6
Bakelite	16×10^6
Mylar	16×10^6
polystyrene	24×10^6
Teflon	16×10^6
glass	120×10^6

Voltage breakdown is a spark, or arc, between the two plates. This is an ionization of the dielectric molecules and the ionization is a virtual short circuit. This will discharge the capacitor and usually destroy the capacitor.

Even the friction of dry snow blowing across an antenna can cause breakdown of transmission lines. Walking across a wool rug in winter can cause static electricity buildup, and an arc can be created by touching a grounded object such as a doorknob.

Lightning is the ultimate breakdown. The air between storm clouds and the ground becomes ionized and huge currents flow, destroying objects in their path. Tales abound of lightning's destructive nature. Lightning protection is important for buildings, antennas, and power transmission lines. A lightning protection device can be a capacitor especially designed to break down in a controlled and harmless manner by shorting out the lightning current.

Large power transmission lines have the top two wires connected directly to the tower's frame, providing a path to ground. These two wires are lightning catchers. They protect the three wires below the top of the tower that actually carry the power.

Example 1.61

An air capacitor has a 0.5 mm plate separation. What is the maximum voltage that can be safely put onto that capacitor?

Solution

From Table 1.6, the breakdown electric field strength is 3×10^6 V/m.

$$\overline{E}_{\max} = 3 \times 10^6 \text{ V/m}$$

$$V_{\max} = \overline{E}_{\max} d = \left(3 \times 10^6 \, \frac{\text{V}}{\text{m}}\right) \left(0.5 \times 10^{-3} \text{ m}\right)$$

$$= 1.5 \times 10^3 \text{ V} \quad (1.5 \text{ kV})$$

PARALLEL CAPACITORS

Figure 1.19 shows two capacitors with identical permittivities and plate spacing and that are connected in parallel.

The capacitance equation is

$$C = \frac{\varepsilon A}{d}$$

The general equation for total capacitance is

$$C_{\text{total}} = C_1 + C_2 + ... + C_n$$

Figure 1.19 Parallel Capacitors

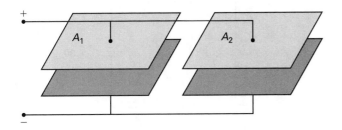

For two identical, parallel capacitors with areas A_1 and A_2, such as those in Fig. 1.19,

$$C_{\text{total}} = \frac{\varepsilon A_1}{d} + \frac{\varepsilon A_2}{d}$$
$$= \frac{\varepsilon}{d}(A_1 + A_2)$$
$$= C_1 + C_2$$

The general parallel capacitor equation is

$$C_{\text{total}} = C_1 + C_2 + ... + C_n \qquad 1.36$$

Example 1.62

In the illustration, three capacitors of 100 pF, 220 Pf, and 470 pF are connected in parallel. What is their total capacitance?

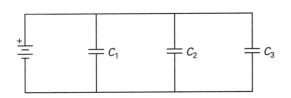

Solution

$$C_{\text{total}} = C_1 + C_2 + C_3$$
$$= 100 \text{ pF} + 220 \text{ pF} + 470 \text{ pF}$$
$$= 790 \text{ pF}$$

SERIES CAPACITORS

Series capacitors, as the name suggests, are capacitors connected in series to a voltage source. Figure 1.20 shows an edge view of two capacitors in series charged by a battery.

For the following explanation of how to simplify series capacitors, assume the capacitors in Fig. 1.20 have the same dielectric, and that their plates have the same area separated by the same distance. The two center plates in Fig. 1.20(a) are connected electrically, so these two plates can be combined as in Fig. 1.20(b). The center plate does not affect the capacitance, so it can be removed as in Fig. 1.20(c). The capacitor now has twice the original spacing between the plates and thus half the capacitance.

$$C_{\text{series}} = \frac{\varepsilon A}{2d} \qquad 1.37$$

The formulas for series capacitors are similar to resistances in parallel. For two capacitors in series, the total capacitance is

$$C_{\text{total}} = \frac{C_1 C_2}{C_1 + C_2} \qquad 1.38$$

The general equation for series capacitors is

$$C_{\text{total}} = \frac{1}{\dfrac{1}{C_1} + \dfrac{1}{C_2} + ... \dfrac{1}{Cn}} \qquad 1.39$$

Or,

$$\frac{1}{C_{\text{total}}} = \frac{1}{C_1} + \frac{1}{C_2} + ... \frac{1}{C_n} \qquad 1.40$$

The voltage on any individual capacitor is

$$V_i = V_{\text{total}} \frac{C_{\text{total}}}{C_i} \qquad 1.41$$

The subscript i identifies the individual capacitor. The charge on any capacitor is

$$Q_i = C_i V_i \qquad 1.42$$

The total charge on the series capacitor string is the same as the charge on any capacitor in the series string.

Figure 1.20 Simplification of Series Capacitors

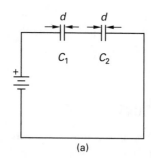

(a)

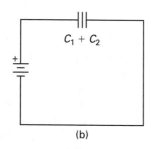

(b)

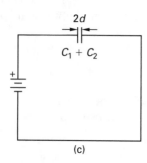

(c)

Example 1.63

What is the total capacitance of the two series capacitors in the illustration?

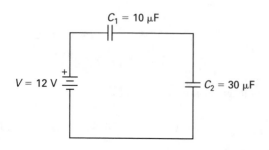

Solution

$$C_{\text{total}} = \frac{C_1 C_2}{C_1 + C_2} = \frac{(10\ \mu\text{F})(30\ \mu\text{F})}{10\ \mu\text{F} + 30\ \mu\text{F}}$$

$$= 7.5\ \mu\text{F}$$

$$\frac{1}{C_{\text{total}}} = \frac{1}{C_1} + \frac{1}{C_2} = \frac{1}{10\ \mu\text{F}} + \frac{1}{30\ \mu\text{F}}$$

$$= \frac{0.1}{1\ \mu\text{F}} + \frac{0.0333}{1\ \mu\text{F}}$$

$$= \frac{0.0333}{1\ \mu\text{F}}$$

$$C_{\text{total}} = \frac{1\ \mu\text{F}}{0.0333}$$

$$= 7.5\ \mu\text{F}$$

Example 1.64

Find the voltage on both capacitors in the illustration.

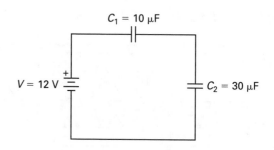

Solution

Insert total capacitance from the previous example to find the voltage on each capacitor.

$$V_1 = V_{\text{total}} \frac{C_{\text{total}}}{C_1}$$

$$= (12\ \text{V}) \left(\frac{7.5\ \mu\text{F}}{10\ \mu\text{F}} \right)$$

$$= 9\ \text{V}$$

$$V_2 = V_{\text{total}} \frac{C_{\text{total}}}{C_2}$$

$$= (12\ \text{V}) \left(\frac{7.5\ \mu\text{F}}{30\ \mu\text{F}} \right)$$

$$= 3\ \text{V}$$

$$V_{\text{total}} = V_1 + V_2 = 9\ \text{V} + 3\ \text{V}$$

$$= 12\ \text{V}$$

$$Q_1 = C_1 V_1 = (10\ \mu\text{F})(9\ \text{V})$$

$$= 90\ \mu\text{C}$$

$$Q_2 = C_2 V_2 = (30\ \mu\text{F})(3\ \text{V})$$

$$= 90\ \mu\text{C}$$

$$Q_{\text{total}} = C_{\text{total}} V_1$$

$$= (7.5\ \mu\text{F})(12\ \text{V})$$

$$= 90\ \mu\text{C}$$

KVL rules apply to all electrical circuits, including capacitors. Note the charge on each capacitor is the same as the total charge. This value can be used to calculate the voltage on each capacitor.

Example 1.65

Find the voltage across each of the four capacitors in the illustration and confirm KVL. Find the total system charge, and the charge and voltage on each capacitor.

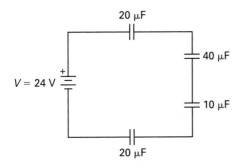

Solution

$$\frac{1}{C_{\text{total}}} = \frac{1}{C_1} + \frac{1}{C_2} + \frac{1}{C_3} + \frac{1}{C_4}$$

$$= \frac{1}{20 \ \mu\text{F}} + \frac{1}{40 \ \mu\text{F}} + \frac{1}{10 \ \mu\text{F}} + \frac{1}{20 \ \mu\text{F}}$$

$$= \frac{0.05}{1 \ \mu\text{F}} + \frac{0.025}{1 \ \mu\text{F}} + \frac{0.1}{1 \ \mu\text{F}} + \frac{0.05}{1 \ \mu\text{F}}$$

$$= \frac{0.225}{1 \ \mu\text{F}}$$

$$C_{\text{total}} = \frac{1 \ \mu\text{F}}{0.225} = 4.4 \ \mu\text{F}$$

$$V_1 = V_{\text{total}} \frac{C_{\text{total}}}{C_1}$$

$$= (24 \ \text{V}) \left(\frac{4.444 \ \mu\text{F}}{20 \ \mu\text{F}} \right)$$

$$= 5.3 \ \text{V}$$

$$V_2 = V_{\text{total}} \frac{C_{\text{total}}}{C_2} = (24 \ \text{V}) \left(\frac{4.444 \ \mu\text{F}}{40 \ \mu\text{F}} \right)$$

$$= 2.7 \ \text{V}$$

$$V_3 = V_{\text{total}} \frac{C_{\text{total}}}{C_3} = (24 \ \text{V}) \left(\frac{4.444 \ \mu\text{F}}{10 \ \mu\text{F}} \right)$$

$$= 10.7 \ \text{V}$$

$$V_4 = V_{\text{total}} \frac{C_{\text{total}}}{C_4} = (24 \ \text{V}) \left(\frac{4.444 \ \mu\text{F}}{20 \ \mu\text{F}} \right)$$

$$= 5.3 \ \text{V}$$

$$V_{\text{total}} = V_1 + V_2 + V_3 + V_4$$

$$= 5.3 \ \text{V} + 2.7 \ \text{V} + 10.7 \ \text{V} + 5.3 \ \text{V}$$

$$= 24 \ \text{V}$$

KVL is satisfied.

$$Q_1 = CV_1 = (20 \ \mu\text{F})(5.3 \ \text{V})$$
$$= 107 \ \mu\text{C}$$
$$Q_2 = CV_2 = (40 \ \mu\text{F})(2.7 \ \text{V})$$
$$= 107 \ \mu\text{C}$$
$$Q_3 = CV_3 = (10 \ \mu\text{F})(10.7 \ \text{V})$$
$$= 107 \ \mu\text{C}$$
$$Q_4 = CV_4 = (20 \ \mu\text{F})(5.3 \ \text{V})$$
$$= 107 \ \mu\text{C}$$
$$Q_{\text{total}} = C_{\text{total}} V_{\text{total}}$$
$$= (4.444 \ \mu\text{F}) \ (24 \ \text{V})$$
$$= 106.7 \ \mu\text{C} \quad (107 \ \mu\text{C})$$

The charge on each capacitor is equal to the charge on the capacitor string.

ENERGY STORAGE IN CAPACITORS

Charged capacitors store energy. The following equation shows the relationship between stored energy, E, capacitance value, and voltage.

$$E = \tfrac{1}{2}CV^2 \qquad\qquad 1.43$$

This equation holds true even for a vacuum capacitor, in which the dielectric is a vacuum with a relative permittivity of $\varepsilon_r = 1.000$. Therefore, the energy must be stored in the electric field. If a dielectric with $\varepsilon_r > 1.000$ is between the plates, the capacitance increases and the stored energy increases proportionally. This means that in these capacitors some energy is also stored in the dielectric. The electric field shifts the atoms in the dielectric, storing energy. This might be a test question.

Example 1.66

A 10 μF capacitor is charged to 12 V. How much energy is stored in the capacitor?

Solution

$$E = \tfrac{1}{2}CV^2$$

$$= \frac{(10 \ \mu\text{F}) \ (12 \ \text{V})^2}{2}$$

$$= 720 \ \mu\text{J}$$

CHARGING AND DISCHARGING A CAPACITOR

Current and voltage sources charge and discharge capacitors. A source supplies electrons charging a capacitor. If the source reduces its amperage or voltage, the electrons will return to the source. The source can change polarity and electrons return to the source and the other plate is charged with electrons. The direction and amplitude of electron or current travel can change, giving the impression that current flows through a capacitor. This will be discussed more in the alternating-current (AC) circuits chapter.

The dielectric is an insulator, so few electrons travel across it. A miniscule leakage current does flow between the plates of any real capacitor. The magnitude of this leakage current depends on the dielectric quality and applied voltage. An ideal capacitor (which exists only in theory) does not have leakage current. Using ideal components in circuit analysis will yield satisfactory results in most cases.

The change in voltage in a capacitor can be found with an alternate form of Eqs. 1.25 and 1.26.

$$Q = CV$$
$$V = \frac{Q}{C}$$
$$\Delta V = \frac{\Delta Q}{C}$$

Since current is electrons per unit time (charge divided by time), dividing both sides of the equation by the change in time Δt, yields the capacitor voltage rate of change as a function of current.

$$\frac{\Delta V}{\Delta t} = \left(\frac{1}{C}\right)\left(\frac{\Delta Q}{\Delta t}\right) = \frac{1}{C}i$$
$$= \frac{i}{C} \qquad 1.44$$

Example 1.67

The current waveform in the illustration is charging a 0.1 μF capacitor. Find the voltage on that capacitor.

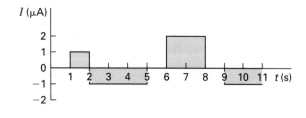

Solution

$$\frac{\Delta V}{\Delta t} = \frac{I}{C}$$

The rate of change in capacitor voltage is proportional to the current into or out of the capacitor. Between 0 s and 1 s, no current flows, so there is no change in voltage. Between 1 s and 2 s, a positive current of 1.0 μA flows. The rate of voltage change is

$$\Delta V = \frac{I\Delta t}{C} = \frac{\left(1.0 \times 10^{-6}\ \text{A}\right)\left(1\ \text{s}\right)}{1 \times 10^{-7}\ \text{F}}$$
$$= 10\ \text{V}$$

Between 2 s and 5 s, the current is $-1.0\ \mu$A, and the voltage change is

$$\Delta V = \frac{I\Delta t}{C}$$
$$= \frac{\left(-1.0 \times 10^{-6}\ \text{A}\right)\left(5\ \text{s} - 2\ \text{s}\right)}{1 \times 10^{-7}\ \text{F}}$$
$$= -30\ \text{V}$$

The voltage changes from 10 V to -20 V between 2 s and 5 s.

Between 5 s and 6 s, no current flows and the voltage remains -20 V. The current is 2.0 μA between 6 s and 8 s, and the voltage change is

$$\Delta V = \frac{I\Delta t}{C}$$
$$= \frac{\left(2.0 \times 10^{-6}\ \text{A}\right)\left(8\ \text{s} - 6\ \text{s}\right)}{1 \times 10^{-7}\ \text{F}}$$
$$= 40\ \text{V}$$

The voltage changes from -20 V to 20 V.

Between 8 s and 9 s, no current flows. The voltage stays at 20 V.

After 9 s, the current is $-1.0\ \mu$A, and the voltage changes at a rate of

$$\frac{\Delta V}{\Delta t} = \frac{I}{C} = \frac{-1.0 \times 10^{-6}\ \text{A}}{1 \times 10^{-7}\ \text{F}}$$
$$= -10\ \text{V/s}$$

The changes in the voltage in the capacitor over time are shown in the following illustration.

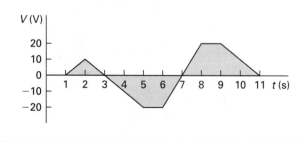

LOW-PASS RESISTOR-CAPACITOR TIME CONSTANT

A circuit with a resistor and capacitor is called an RC circuit. The resistor will limit the current flow into or out of a capacitor and thus limit the charging rate. So RC circuits can be used for timing and frequency filters. Frequency filters can either pass low frequencies and reduce, or attenuate, high frequencies—these types are called low-pass (LP) filters—or they can pass high frequencies and reduce low frequencies—these types are called high-pass (HP) filters.

RC circuits can also be band-pass (BP) and band-reject (BR) filters. BP filters pass only frequencies in a small band and reduce frequencies above and below the desired frequency. An example of the use of a BP filter is selecting stations in a radio receiver. BR filters reject frequencies in a certain band. An example of the use of a BR filter is removing 60 Hz noise from a signal.

An RC timing circuit is shown in Fig. 1.21. The circuit has a pulse generator driving the resistor-capacitor series combination. The current into the capacitor is not constant.

$$I = \frac{V - V_C}{R} \qquad 1.45$$

At time $t = 0$ s, the capacitor does not have any charge and so its voltage is 0 V. Over time, current flows into the capacitor, which stores charge and gains voltage. The voltage difference between the generator voltage, V_{in}, and the capacitor voltage, V_C, is reduced, and with this change the voltage and current through the resistor are also reduced. This in turn reduces the charge rate.

The example in the previous section showed a capacitor charging and discharging linearly with constant current. In this example, the capacitor voltage does not increase linearly, but rather exponentially. The equation that describes this process is a differential equation, but students only need to know the differential equation solution, shown in Eq. 1.46. i is the instantaneous current through the resistor and into the capacitor. τ is the RC time constant. The time constant is the product of resistance and capacitance ($\tau = RC$).

$$i = \frac{V_{\text{in}}}{R} e^{-t/RC} = \frac{V_{\text{in}}}{R} e^{-t/\tau} \qquad 1.46$$

The unit of the RC time constant is time in seconds. The exponent has time units in the numerator and in the denominator; so the exponent is unitless. A larger resistor slows the charging rate, and a larger capacitor takes longer to charge. In this way, changing the resistor or capacitor controls the charge rate.

As seen in Fig. 1.22, the current in an RC circuit decreases exponentially with time and approaches 0 μA. V is a step function.

Figure 1.21 RC Charging

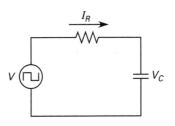

Figure 1.22 Current over Time in a Resistor-Capacitor Circuit

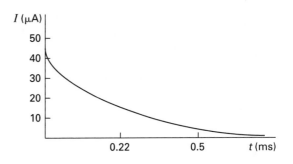

This type of circuit is used for LP filters, timing circuits, and crude integrators. Operational amplifiers (op amps) make much better integrators at the expense of increased complexity. Op amps and integrators are covered in Ch. 4.

The capacitor voltage is

$$\begin{aligned} V_c &= V_{\text{in}} \left(1 - e^{-t/RC} \right) \\ &= V_{\text{in}} \left(1 - e^{-t/\tau} \right) \qquad 1.47 \end{aligned}$$

Example 1.68

An RC circuit like the one shown in Fig. 1.21 has the following parameters: $R = 220$ kΩ, $C = 1.0$ μF, and a pulse generator supplying a 10 V step function. Find the current at 0.0 s and 22 ms.

Solution

$$\begin{aligned} RC &= \left(220 \times 10^3 \ \Omega \right) \left(0.1 \times 10^{-6} \ \text{F} \right) \\ &= 22 \times 10^{-3} \ \text{s} \quad (22 \ \text{ms}) \end{aligned}$$

At a time of 0.0 s,

$$\begin{aligned} i_c &= \frac{V_{\text{in}}}{R} e^{-t/RC} \\ &= \left(\frac{10 \ \text{V}}{0.220 \ \text{M}\Omega} \right) e^{-0 \ \text{s}/22 \ \text{ms}} \\ &= 45.45 \ \mu\text{A} \end{aligned}$$

At a time of 0.22 ms,

$$i_c = \frac{V_{\text{in}}}{R} e^{-t/RC}$$

$$= \left(\frac{10 \text{ V}}{0.220 \text{ M}\Omega} \right) e^{-22 \text{ ms}/22 \text{ ms}}$$

$$= (45.45 \ \mu\text{A}) e^{-1}$$

$$= 16.72 \ \mu\text{A}$$

Figure 1.22 shows the capacitor current.

Example 1.69

Find voltage across the 200 nF capacitor in the illustration at 0.0 μs, 250 μs, and 500 μs. The source is a 10 V step function.

Solution

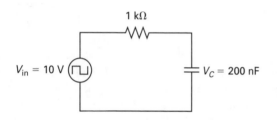

$$RC = \left(1.0 \times 10^3 \ \Omega \right) \left(200 \times 10^{-9} \text{ F} \right)$$

$$= 200 \times 10^{-6} \text{ s} \quad (200 \ \mu\text{s})$$

At a time of 0 μs,

$$V_C = V_{\text{in}} \left(1 - e^{-t/RC} \right)$$

$$= (10 \text{ V}) \left(1 - e^{-0 \ \mu\text{s}/200 \ \mu\text{s}} \right)$$

$$= (10 \text{ V}) \left(1 - e^0 \right)$$

$$= (10 \text{ V}) (1 - 1)$$

$$= 0 \text{ V}$$

At a time of 250 μs,

$$V_C = V_{\text{in}} \left(1 - e^{-t/RC} \right)$$

$$= (10 \text{ V}) \left(1 - e^{-250 \ \mu\text{s}/200 \ \mu\text{s}} \right)$$

$$= (10 \text{ V}) \left(1 - e^{-1.25} \right)$$

$$= (10 \text{ V}) (1 - 0.2865)$$

$$= 7.1 \text{ V}$$

Figure 1.23 HP Filter

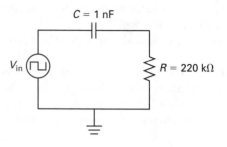

At a time of 500 μs,

$$V_C = V_{\text{in}} \left(1 - e^{-t/RC} \right)$$

$$= (10 \text{ V}) \left(1 - e^{-500 \ \mu\text{s}/200 \ \mu\text{s}} \right)$$

$$= (10 \text{ V}) \left(1 - e^{-2.5} \right)$$

$$= (10 \text{ V}) (1 - 0.082)$$

$$= 9.2 \text{ V}$$

The plot of capacitor voltage over time is shown in the following illustration. As time becomes much greater than τ, the RC product, V_C approaches $V_{\text{in}} = 10$ V.

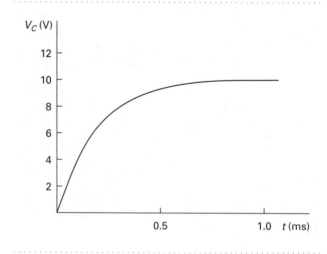

HIGH-PASS RESISTOR-CAPACITOR TIME CONSTANT

A high-frequency RC filter can be made by connecting a resistor and capacitor as shown in Fig. 1.23.

The voltage across the resistor is given by the following equation.

$$V_R = V_{\text{in}} e^{-t/RC} = V_{\text{in}} e^{-t/\tau} \qquad 1.48$$

At a time of 0 s, the capacitor is in a discharged state, and the capacitor voltage is 0 V. By KVL, the resistor voltage equals the source voltage at a time of 0 s. Over time, current flowing through the resistor charges the capacitor and increases the capacitor voltage. The current and resistor voltage must then decrease by KVL.

Example 1.70

Using the circuit in Fig. 1.23, find the resistor voltage at 0 μs, at 250 μs, and at 500 μs. The voltage source is a signal source with a repeating square wave of 500 μs at 10 V and 500 μs at 0 V.

Solution

$$\tau = RC$$
$$= (220 \text{ k}\Omega) \left(1000 \; \frac{\Omega}{\text{k}\Omega}\right) (1.0 \times 10^{-9} \text{ F})$$
$$= 0.00022 \text{ s} \quad (220 \; \mu\text{s})$$

For a time of 0 μs,

$$V_R = V_{\text{in}} e^{-t/RC}$$
$$= (10 \text{ V}) \, e^{-0 \; \mu\text{s}/220 \; \mu\text{s}}$$
$$= 10 \text{ V}$$

At 250 μs,

$$V_R = V_{\text{in}} e^{-t/RC}$$
$$= (10 \text{ V}) \, e^{-250 \; \mu\text{s}/220 \; \mu\text{s}}$$
$$= 3.210 \text{ V}$$

At 500 μs,

$$V_R = V_{\text{in}} e^{-t/RC}$$
$$= (10 \text{ V}) \, e^{-500 \; \mu\text{s}/220 \; \mu\text{s}}$$
$$= 1.030 \text{ V}$$

The voltage versus time plot is shown in the following illustration.

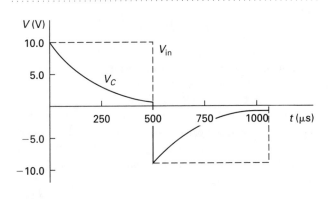

MAGNETISM

A permanent magnet creates a magnetic field. The magnetic flux lines are said to emanate from the north pole and terminate at the south pole. Flux lines, as well as the north and south labels, are a useful way to imagine a magnetic field, but they should not to be taken too literally. The flux lines indicate the general orientation of the magnetic field but do not mark distinct pathways. The north and south labels for magnet poles are based on the earth's magnetic field.

A simple experiment reveals the magnetic flux lines of a permanent magnet. A sheet of paper is placed over a bar magnet. Iron filings are placed on the paper. The iron filings orient themselves along the magnetic flux lines, similar to the lines shown in Fig. 1.24.

Figure 1.24 Permanent Magnetic Flux Lines

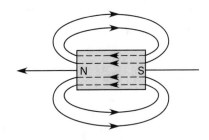

Magnetism Created by an Electric Current

Electric current is electrons in motion. The electrons are electrical charges. Charges in motion create a magnetic field. The magnetic field direction can be found by the right hand rule, shown illustrated in Fig. 1.25. If the right hand thumb is pointing in the direction of conventional current, the fingers will point in the direction of the magnetic field. Figure 1.26 shows an end view of the magnetic flux created by electrical current.

Figure 1.25 Right Hand Rule for Magnetic Fields

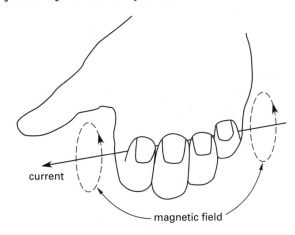

MAGNETIC FLUX DENSITY

Magnetic flux has units of webers (Wb), and is represented by the symbol Φ. Of greater use to electrical engineers is the magnetic flux density, B, which is magnetic flux (in webers) per unit area (in square meters). The unit weber/meters^2 is equivalent to the Tesla (T).

$$B = \frac{\Phi}{A} \qquad 1.49$$

Figure 1.26 Magnetic Fields Created by Electrical Current (End View)

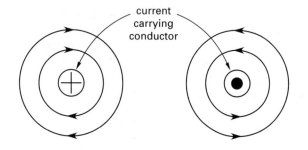

The flux density created by a current in a long wire in free space is given by the following equation. r is the distance from the wire to the point where the measurement is made. In this case, "long" means the distance r is much less than the length of the wire, and the wire diameter is much less than the distance r to the measurement point. These approximations simplify the equation.

$$B = \left(2 \times 10^{-7} \, \frac{\text{Wb}}{\text{A·m}} \right) \frac{I}{r} \qquad 1.50$$

Example 1.71

A bar magnet with an area of 1 cm × 1 cm has a magnetic flux of 30×10^{-6} Wb. What is the flux density?

Solution

$$B = \frac{\Phi}{A} = \frac{30 \times 10^{-6} \, \text{Wb}}{(0.01 \, \text{m})^2}$$
$$= 30 \times 10^{-2} \, \text{T} \quad (0.3 \, \text{T})$$

Example 1.72

A long wire is carrying a current of 2.0 A. What is the magnetic field density 10 cm from the wire?

Solution

$$B = \left(2 \times 10^{-7} \right) \frac{I}{r}$$
$$= \left(2 \times 10^{-7} \, \frac{\text{Wb}}{\text{A·m}} \right) \left(\frac{2.0 \, \text{A}}{0.1 \, \text{m}} \right)$$
$$= 4 \times 10^{-6} \, \text{T}$$

Note the flux density for a point source falls off as the inverse square of the distance; the flux density for the line source (e.g., a current-carrying wire) falls off as the inverse distance.

MAGNETIC FIELD INTENSITY

The intensity of magnetic fields can be increased by winding the conducting wire into a coil. Often a coil is wrapped around a bar of magnetic material such as iron, cobalt, or nickel. Passing a current through the coil creates a magnetic field more intense than if the wire were not coiled. The parameter of field intensity is different than flux or flux density. The field intensity has units of ampere-turns per meter (A·turn/m) and is given the symbol H. N is the number of turns of the wire, I is the current in amperes, and l is the length in meters of the bar around which the coil is wrapped.

$$H = \frac{NI}{l} \qquad 1.51$$

Example 1.73

A 150-turn coil is carrying a current of 3.0 A. The coil is wound around an iron toroid (a torus or ring) 20 cm in circumference. What is the magnetic field intensity?

Solution

$$H = \frac{NI}{l} = \frac{(150 \, \text{turns})(3.0 \, \text{A})}{0.2 \, \text{m}}$$
$$= 2250 \, \text{A·turn/m}$$

PERMEABILITY

Substances have a quantity called resistivity which describes the materials ability to conduct electricity. Similarly, materials have a quantity describing their ability to carry magnetic flux. This quantity is called permeability and is given the symbol μ. Remember permittivity in capacitors ties charge and distance into SI units. Permeability relates magnetic quantities to meters and current. The relationship between magnetic field intensity and magnetic flux density is

$$B = \mu H = \mu \frac{NI}{l} \qquad 1.52$$

The permeability of air is $4\pi \times 10^{-7}$ Wb/A·turn·m and is given the symbol μ_0. Most materials have permeabilities close to the permeability of air. Magnetic materials such as iron, cobalt, nickel, and gadolinium have permeabilities much higher than this. These magnetic materials are described as ferromagnetic, from the Latin word *ferrum*, meaning iron. Their permeabilities are found by multiplying the material's relative permeability, μ_r, by the permeability of air.

$$\mu = \mu_r \mu_0 \qquad 1.53$$

The relative permeability of most substances is about 1.00, but the relative permeability for ferromagnetic materials is several hundred. For ferromagnetic materials, Eq. 1.53 must be slightly modified.

$$B = \mu H = \mu_r \mu_0 H \qquad 1.54$$

Example 1.74

A 220-turn coil is wound on an iron toroid with an 11 cm circumference. The relative permeability of the iron is 1100. What is the flux density if the coil has a current of 1.5 A?

Solution

$$B = \mu H = \mu_r \mu_0 \frac{NI}{l}$$

$$= (1100) \left(4\pi \times 10^{-7} \; \frac{\mathrm{Wb}}{\frac{\mathrm{A \cdot turn}}{\mathrm{m}}} \right)$$

$$\times \left(\frac{(220 \text{ turns}) (1.5 \text{ A})}{0.11 \text{ m}} \right)$$

$$= 4.1 \; \frac{\mathrm{Wb}}{\mathrm{m^2}} \quad (4.1 \text{ T})$$

MAGNETIC DOMAINS

A magnetic material can be thought of as comprising many small magnets, called *domains*. When the material is demagnetized, the domains have a random orientation, as shown in Fig. 1.27. The arrows show the magnetization direction.

Figure 1.27 *Randomly Oriented Domains*

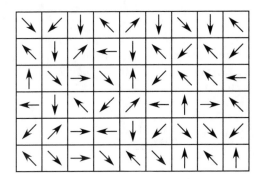

If exposed to an external magnetic field, the domains align themselves with the external field as shown in Fig. 1.28. The external field can be generated by a permanent magnet or by an electrical current.

Figure 1.28 *Partially Aligned Magnetic Domains*

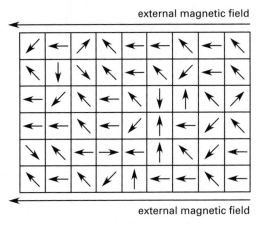

Note not all the domains in Fig. 1.28 are completely aligned with the external magnetic field. As the external field increases in intensity, more domains will align themselves with the field. If the external magnetic field is removed, most of the domains will retain their magnetized orientation, but some will take a random orientation.

Figure 1.29 plots the general orientation of domains under the influence of varying magnetic intensities in a hypothetical situation. An unmagnetized ferromagnetic material with randomly aligned domains and no magnetizing force is at point W. The material is then subjected to a magnetizing force with a field intensity, H. As the field intensity increases, the domains begin to align themselves with the increasing magnetizing force and the magnetic flux density, B, increases. The magnetic flux density increases linearly from point W to point X. Beyond point X most of the domains have been aligned and there are few unaligned domains remaining. Eventually, most of the domains are aligned and increasing field intensity has less effect. This gives rise to a nonlinear curve from point X to point Y. When the field intensity is reduced, many domains remain aligned. The flux density does not return to 0 T when the field intensity is reduced to 0 A·turn/m, creating a permanent magnet.

The slope of the *B-H* curve is the material's permeability.

$$\mu = \frac{B}{H} \approx \frac{\Delta B}{\Delta H} \qquad 1.55$$

Example 1.75

What is the relative permeability of the linear portion of the curve in Fig. 1.29?

Figure 1.29 B-H Curve

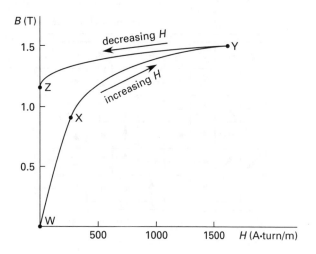

Solution

The linear portion is between points W and X.

$$\mu = \frac{B}{H} = \frac{\Delta B}{\Delta H}$$

$$= \frac{0.9 \text{ T}}{300 \ \frac{\text{A·turn}}{\text{m}}}$$

$$= 0.003 \text{ Wb/A·turn·m}$$

$$\mu_r = \frac{\Delta \mu}{\mu_0}$$

$$= \frac{0.003 \ \dfrac{\text{Wb}}{\dfrac{\text{A·turn}}{\text{m}}}}{4\pi \times 10^{-7} \ \dfrac{\text{Wb}}{\dfrac{\text{A·turn}}{\text{m}}}}$$

$$= 2387 \quad (2400)$$

INDUCED VOLTAGE

Faraday's law states that a changing magnetic field will induce a voltage in a conductor. The induced voltage is calculated with the following equation. N is the number of turns in a coil, and Φ is the magnetic flux in webers.

$$V = -N\frac{\Delta \Phi}{\Delta t} \qquad 1.56$$

As can be seen from the equation, a wire coil increases the induced voltage. Many texts incorrectly do not include the minus sign.

Faraday's law is the basis of electrical generators. Coils of wire are driven by some mechanical force (usually an engine, water, or steam) through a magnetic field, generating a voltage.

Energy is stored in a magnetic field. A change in electrical current changes the stored energy. Joseph Henry found that a current change in an inductor would return the stored energy to the conductor. This energy is a voltage, sometimes called a "back emf" (for electromotive force). This property is called inductance and given units of henrys (H) and the symbol L. The voltage change is

$$e = -L\frac{\Delta I}{\Delta t} \qquad 1.57$$

Example 1.76

A coil has 120 turns and is in a magnetic field with a magnetic flux rate of change of 1.6×10^{-3} Wb in 8 ms. What is the induced voltage?

Solution

$$e = -N\frac{\Delta \Phi}{\Delta t} = -(120 \text{ turns})\left(\frac{1.6 \times 10^{-3} \text{ Wb}}{8 \text{ ms}}\right)$$

$$= -24 \text{ V}$$

Example 1.77

A coil has an inductance of 40 mH and a current changing at a rate of 120 mA in 6 μs. Find the induced voltage.

Solution

$$e = -L\frac{\Delta I}{\Delta t} = -(40 \text{ mH})\left(\frac{120 \text{ mA}}{6 \text{ }\mu\text{s}}\right)$$

$$= -800 \text{ V}$$

SERIES INDUCTORS

Inductors in series add their individual inductances in a similar way that resistors in series add resistances. This assumes the magnetic fields from each inductor do not affect each other.

$$L_{\text{total}} = L_1 + L_2 + ... + L_i \qquad 1.58$$

Example 1.78

For the inductors in the illustration, find the total inductance. Assume the inductors' magnetic fields do not affect each other.

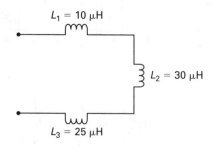

Solution

$$L_{\text{total}} = L_1 + L_2 + L_3$$
$$= 10 \ \mu\text{H} + 30 \ \mu\text{H} + 25 \ \mu\text{H}$$
$$= 65 \ \mu\text{H}$$

ENERGY STORED IN AN INDUCTOR

Current creates a magnetic field and energy is stored in that magnetic field. Any charged particle in motion creates a magnetic field and thus stores energy. The equation for energy stored in an inductor's magnetic field is

$$W = \frac{LI^2}{2} \qquad 1.59$$

Note the similarity to the equation for energy stored in a capacitor. Voltage on a capacitor stores energy in the capacitor's electric field.

Example 1.79

A current of 1.5 A flows in a 0.5 H inductor. Find the stored energy.

Solution

$$W = \frac{LI^2}{2} = \frac{(0.5 \ \text{H}) (1.5 \ \text{A})^2}{2}$$
$$= 0.5625 \ \text{J}$$

INDUCTIVE TRANSIENTS

An inductor-resistor combination creates a time constant, just as a resistor-capacitor combination does. Figure 1.30 shows a simple inductor-resistor timing circuit.

The inductor-resistor time constant is

$$\tau = \frac{L}{R} \qquad 1.60$$

Figure 1.30 Inductive Transients

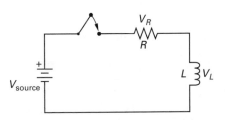

The voltage across the inductor is

$$V_L = V_{\text{source}} e^{-t/\tau} \qquad 1.61$$

Initially, the inductor does not allow any current flow, and the inductor voltage is equal to the source voltage. This is because it takes time for the current to build up the magnetic field. When the switch opens, current no longer flows and the magnetic field collapses, transferring the energy stored in the magnetic field into a voltage transient. When the inductive transient dies away, the inductor voltage reduces to 0 V.

Inductance is used by older automobile ignition systems to generate a spark and ignite the gasoline-air mixture in the cylinders. The spark is generated by breaking a current through an ignition coil by opening a switch or points. (Points used to be closed and opened by the engine camshaft, but this ignition type has been replaced by electronic ignitions.) About 350 V are generated. A transformer steps up the voltage to over 10 kV and an arc jumps the spark plug gap inside the engine cylinder, igniting the gasoline vapors. Transformers will be discussed in Ch. 2. A simplified automobile ignition system is illustrated in Fig. 1.31.

Figure 1.31 Automobile Ignition System

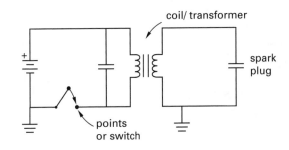

Example 1.80

A circuit contains a 130 μH inductor with a wire resistance of 22 Ω. The source voltage is 12 V. Find the time constant, and find the voltages across the inductor at times of 0 μs, 4 μs, 10 μs, and 100 μs.

Solution

$$\tau = \frac{L}{R} = \frac{130 \ \mu\text{H}}{22 \ \Omega}$$
$$= 5.9 \ \mu\text{s}$$

At a time of 0 μs,

$$V_L = V_{\text{source}} e^{-t/\tau}$$
$$= (12 \ \text{V}) \, e^{-0 \ \mu\text{s}/5.9 \ \mu\text{s}}$$
$$= (12 \ \text{V}) \, e^{-0}$$
$$= (12 \ \text{V}) \, (1.0)$$
$$= 12 \ \text{V}$$

At a time of 4 μs,

$$V_L = V_{source}e^{-t/\tau}$$
$$= (12 \text{ V}) \, e^{-4 \, \mu s/5.9 \, \mu s}$$
$$= (12 \text{ V}) \, e^{-0.67}$$
$$= (12 \text{ V}) \, (0.5117)$$
$$= 6.1 \text{ V}$$

At a time of 10 μs,

$$V_L = V_{source}e^{-t/\tau}$$
$$= (12 \text{ V}) \, e^{-10 \, \mu s/5.9 \, \mu s}$$
$$= (12 \text{ V}) \, e^{-0.67}$$
$$= (12 \text{ V}) \, (0.1836)$$
$$= 2.2 \text{ V}$$

At a time of 100 μs,

$$V_L = V_{source}e^{-t/\tau}$$
$$= (12 \text{ V}) \, e^{-100 \, \mu s/5.9 \, \mu s}$$
$$= (12 \text{ V}) \, e^{-16.95}$$
$$\approx (12 \text{ V}) \, (0.0)$$
$$= 0.0 \text{ V}$$

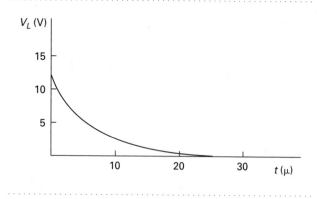

HIGH-INDUCTANCE COILS

Inductors are used to create magnetic fields. Inductance is increased by coiling a wire. Inductance can be further increased by placing a ferromagnetic material near or preferably within the coil, much as capacitance was increased by placing a dielectric between the capacitor plates.

Most materials do not affect inductors and their associated magnetic fields appreciably. However ferromagnetic metals such as iron, nickel, cobalt, and gadolinium greatly increase inductance and magnetic fields.

The increase in relative permeability ranges from several hundred to three thousand. Very strong permanent magnets are made from alloys of ferromagnetic metals and aluminum. Magnets in electric motors and relays use ferromagnetic metals to increase the power of their magnets. Power transformers use iron alloys to increase their internal magnetic fields and greatly increase their efficiencies. Transformers will be studied in the AC chapter.

PRACTICE PROBLEMS

1. A series circuit has three resistors of 10 Ω, 20 Ω, and 30 Ω. The circuit is driven by a 12 V battery.

(a) What is the total resistance?

- (A) 0.18 Ω
- (B) 5.5 Ω
- (C) 30 Ω
- (D) 60 Ω

(b) What is the total current?

- (A) 0.2 A
- (B) 0.6 A
- (C) 1.2 A
- (D) 2.2 A

(c) What is the voltage across each of the 10 Ω, 20 Ω, and 30 Ω resistors, respectively?

- (A) 2 V, 4 V, 6 V
- (B) 4 V, 4 V, 4 V
- (C) 6 V, 4 V, 2 V
- (D) 12 V, 12 V, 12V

(d) What is the current through each of the 10 Ω, 20 Ω, and 30 Ω resistors, respectively?

- (A) 0.2 A, 0.2 A, 0.2 A
- (B) 0.4 A, 0.6 A, 1.2 A
- (C) 1.2 A, 0.6 A, 0.4 A
- (D) 1.2 A, 1.2 A, 1.2 A

(e) What is the total power dissipated in the circuit?

- (A) 0.20 W
- (B) 2.4 W
- (C) 24 W
- (D) 144 W

(f) What is the power dissipated in each of the 10 Ω, 20 Ω, and 30 Ω resistors, respectively?

- (A) 0.4 W, 0.8 W, 1.2 W
- (B) 1.2 W, 0.6 W, 0.4 W
- (C) 12 W, 12 W, 12 W
- (D) 14.4 W, 7.2 W, 4.8 W

2. Given a parallel circuit with two resistors of 10 Ω and 15 Ω. The circuit is driven by a 12 V battery.

(a) What is most nearly the total resistance?

    (A) −5 Ω
    (B) 0.2 Ω
    (C) 6 Ω
    (D) 30 Ω

(b) What is most nearly the total current?

    (A) 2 A
    (B) 6 A
    (C) 10 A
    (D) 70 A

(c) What is most nearly the voltage across each resistor?

    (A) 4.8 V and 7.2 V
    (B) 6.0 V and 6.0 V
    (C) 12 V and 12 V
    (D) 24 V and 24 V

(d) What is most nearly the current through the 10 Ω resistor and 15 Ω resistor, respectively?

    (A) 0.0 A, 0.0 A
    (B) 0.8 A, 1.2 A
    (C) 1.2 A, 0.8 A
    (D) 2.0 A, 4.0 A

(e) What is most nearly the total power dissipated?

    (A) 2.0 W
    (B) 9.9 W
    (C) 14 W
    (D) 24 W

(f) What is most nearly the power dissipated in each resistor?

    (A) $P_{R_{10\ \Omega}} = 9.6$ W, $P_{R_{15\ \Omega}} = 14$ W
    (B) $P_{R_{10\ \Omega}} = 14$ W, $P_{R_{15\ \Omega}} = 9.6$ W
    (C) $P_{R_{10\ \Omega}} = 14$ W, $P_{R_{15\ \Omega}} = 14$ W
    (D) $P_{R_{10\ \Omega}} = 24$ W, $P_{R_{15\ \Omega}} = 14$ W

3. A parallel circuit has three resistors of 12 Ω, 15 Ω and 20 Ω. The circuit is driven by a 30 V battery.

(a) What is most nearly the total resistance?

    (A) 5.0 Ω
    (B) 6.7 Ω
    (C) 7.5 Ω
    (D) 47 Ω

(b) What is most nearly the total current?

    (A) 2 A
    (B) 3 A
    (C) 6 A
    (D) 200 A

(c) What is most nearly the voltage across each resistor?

    (A) 7.7 V
    (B) 15 V
    (C) 20 V
    (D) 30 V

(d) What is most nearly the current through the 12 Ω, 15 Ω, and 20 Ω resistors, respectively?

    (A) 1.5 A, 2.0 A, 2.5 A
    (B) 2.5 A, 2.0 A, 1.5 A
    (C) 6.0 A, 6.0 A, 6.0 A
    (D) 150 A, 150 A, 150 A

(e) What is most nearly the total power dissipated?

    (A) 5.0 W
    (B) 6.0 W
    (C) 150 W
    (D) 180 W

(f) What is most nearly the power dissipated in each resistor?

    (A) $P_{R_{12\ \Omega}} = 2.5$ W, $P_{R_{15\ \Omega}} = 2.0$ W, $P_{R_{20\ \Omega}} = 1.5$ W

    (B) $P_{R_{12\ \Omega}} = 45$ W, $P_{R_{15\ \Omega}} = 60$ W, $P_{R_{20\ \Omega}} = 75$ W

    (C) $P_{R_{12\ \Omega}} = 75$ W, $P_{R_{15\ \Omega}} = 60$ W, $P_{R_{20\ \Omega}} = 45$ W

    (D) $P_{R_{12\ \Omega}} = 180$ W, $P_{R_{15\ \Omega}} = 180$ W, $P_{R_{20\ \Omega}} = 180$ W

4. A circuit has four parallel resistors of 100 Ω each. What is most nearly the total resistance?

    (A) 15 Ω
    (B) 25 Ω
    (C) 50 Ω
    (D) 400 Ω

5. A circuit has three parallel connected capacitors of 1 μF, 2 μF and 4 μF. What is most nearly the total parallel capacitance value?

(A) 0.57 μF
(B) 1.8 μF
(C) 7.0 μF
(D) 21 μF

6. How much energy is stored in a 100 μF capacitor charged to 200 V?

(A) 1 J
(B) 2 J
(C) 4 J
(D) 4×10^6 J

7. A 5 nF capacitor and a 10 nF capacitor are connected in series. What is most nearly the total capacitance?

(A) 3.3 nF
(B) 7.5 nF
(C) 15 nF
(D) 50 nF

8. Four capacitors, each of 10 pF, are connected in series. What is most nearly the total capacitance?

(A) 2.5 pF
(B) 40 pF
(C) 10 000 pF
(D) 2.5 F

9. Three capacitors of 1.0 μF, 3.0 μF, and 5.0 μF are connected in series. What is most nearly their total capacitance?

(A) 0.65 μF
(B) 1.5 μF
(C) 9.0 μF
(D) 15 μF

10. Two series capacitors of 10 μF each have a 9 V battery connected to the series combination. What is most nearly the voltage across each capacitor?

(A) 0.0 V, 0.0 V
(B) 4.5 V, 4.5 V
(C) 9.0 V, 9.0 V
(D) 18 V, 18 V

11. Two series capacitors have values of $C_1 = 10$ μF and $C_2 = 20$ μF, and there is a 9 V battery connected to the series combination. What is most nearly the voltage across each capacitor?

(A) 3.0 V across C_1, 6.0 V across C_2
(B) 4.5 V across C_1, 4.5 V across C_2
(C) 6.0 V across C_1, 3.0 V across C_2
(D) 9.0 V across C_1, 9.0 V across C_2

12. An RC integrator (LP filter) has values of $C = 1.0$ μF and $R = 2\,k\Omega$. At $t = 0$ ms, a 15 V source is switched onto the integrator. What is most nearly the current at $t = 4$ ms?

(A) 1.0 mA
(B) 2.8 mA
(C) 7.5 mA
(D) 15 mA

13. A 2.2 μF capacitor has no initial charge. At $t = 0$ ms, a 0.5 mA current begins charging the capacitor. What is most nearly the voltage at $t = 10$ ms?

(A) 5.0 μV
(B) 2.3 V
(C) 23 V
(D) 230 V

14. An RC differentiator (HP filter) with values of $R = 10$ kΩ and $C = 1.0$ μF is driven by a 10 V step at $t = 0$ ms. What is most nearly the voltage across the resistor at $t = 15$ ms?

(A) 2.2 V
(B) 9.8 V
(C) 9.9 V
(D) 45 V

15. A coil of 50 turns is in a magnetic field that changes at 1.2 Wb/s. What is most nearly the induced voltage?

(A) -60 V
(B) 40 V
(C) 50 V
(D) 60 V

16. A coil of 70 turns is in a magnetic field that changes at -4.5×10^{-3} Wb in 30 ms. What is most nearly the induced voltage?

(A) -11 V
(B) 0.011 V
(C) 11 V
(D) 320 V

17. Three inductors have values of 10 mH, 20 mH, and 5 mH in series. Assume each inductor is independent of the others. What is most nearly the total inductance?

(A) 3 mH
(B) 10 mH
(C) 40 mH
(D) 1000 mH

18. A 200 mA current flows through a 220 mH inductor. How much energy is stored in the inductor's magnetic field?

(A) 0.0044 J
(B) 0.0088 J
(C) 0.022 J
(D) 4.4×10^6 J

19. A series LR circuit has values of $R = 12\ \Omega$ and $L = 360\ \mu H$. What is the time constant?

(A) 0.33 μs
(B) 30 μs
(C) 30 s
(D) 33 000 s

20. Using the results of the previous problem, what is most nearly the voltage across the inductor at 20 μs if the source is 24 V?

(A) 0.0 V
(B) 12 V
(C) 24 V
(D) 47 V

21. What is most nearly the induced voltage if the current through a coil of 0.4 H changes from 0.5 A to 0.1 A in 10 ms?

(A) −16 V
(B) 4.0 V
(C) 16 V
(D) 20 V

22. A nichrome wire is 4.0 km long with a diameter of 1.0 mm. The resistivity of nichrome is $115 \times 10^{-8}\ \Omega \cdot m$. What is most nearly the wire's resistance?

(A) 5.9 Ω
(B) 1500 Ω
(C) 5900 Ω
(D) 18 000 Ω

23. An AWG no. 22 copper wire 10 000 ft long has an area of 642.40 CM. The resistivity of copper is 10.37 CM·Ω/ft. What is most nearly its resistance?

(A) 0.25 Ω
(B) 51 Ω
(C) 110 Ω
(D) 160 Ω

24. A long copper wire has a resistance of 10.00 Ω at 20°C. The temperature coefficient of copper is 3.93×10^{-3} 1/°C. What is most nearly its resistance at 120°C?

(A) 3.9 Ω
(B) 4.7 Ω
(C) 14 Ω
(D) 15 Ω

25. Two point charges of +30 μC and +40 μC are separated by 12 cm. What is most nearly the force between them? Is this force attractive or repulsive?

(A) 8.3×10^{-8} N attractive
(B) 90 N repulsive
(C) 750 N repulsive
(D) 750 N attractive

26. A circular iron rod 1.0 cm in diameter carries a magnetic flux of 22×10^{-6} Wb. What is most nearly the flux density in the rod?

(A) 7.0×10^{-4} T
(B) 7.0×10^{-2} T
(C) 0.28 T
(D) 0.88 T

27. A long wire carries a current of 120 mA. What is most nearly the magnetic field 5.0 cm from the wire?

(A) 4.8×10^{-7} T
(B) 2.4 T
(C) 4.8 T
(D) 23×10^7 T

SOLUTIONS

1. (a) The series resistance is the sum of the resistors.

$$R_{\text{total}} = R_1 + R_2 + R_3$$
$$= 10\ \Omega + 20\ \Omega + 30\ \Omega$$
$$= 60\ \Omega$$

Note: Option (A), the conductance of a parallel configuration, and option (B), the resistance of a parallel configuration, are both wrong.

Answer is D.

(b) The total current is found by using Ohm's law.

$$I_{\text{total}} = \frac{V}{R_{\text{total}}} = \frac{12 \text{ V}}{60 \text{ }\Omega}$$
$$= 0.2 \text{ A}$$

Answer is A.

(c) There are two methods of finding the voltage across a resistor in a series configuration.

$$V_{R_x} = \frac{R_x}{R_{\text{total}}} V_{\text{source}}$$

$$V_{R_{10}} = \left(\frac{10 \text{ }\Omega}{60 \text{ }\Omega}\right)(12 \text{ V}) = 2 \text{ V}$$

$$V_{R_{20}} = \left(\frac{20 \text{ }\Omega}{60 \text{ }\Omega}\right)(12 \text{ V}) = 4 \text{ V}$$

$$V_{R_{30}} = \left(\frac{30 \text{ }\Omega}{60 \text{ }\Omega}\right)(12 \text{ V}) = 6 \text{ V}$$

The alternative method is

$$V_{R_x} = I_{\text{total}} R_x$$
$$V_{R_{10}} = (0.2 \text{ A})(10 \text{ }\Omega) = 2 \text{ V}$$
$$V_{R_{20}} = (0.2 \text{ A})(20 \text{ }\Omega) = 4 \text{ V}$$
$$V_{R_{30}} = (0.2 \text{ A})(30 \text{ }\Omega) = 6 \text{ V}$$

The sum of the resistor voltages in a series circuit is equal to the source voltage. This is KVL.

Answer is A.

(d) The same current flows through all the series resistors in this circuit, and it is the same as total current, 0.2 A.

Answer is A.

(e) Power is determined by multiplying the source voltage by the total current.

$$P_{\text{total}} = VI_{\text{total}}$$
$$= (12 \text{ V})(0.2 \text{ A})$$
$$= 2.4 \text{ W}$$

Or,

$$P_{\text{total}} = \frac{V^2}{R_{\text{total}}} = \frac{(12 \text{ V})^2}{60 \text{ }\Omega}$$
$$= 2.4 \text{ W}$$

Or,

$$P_{\text{total}} = I_{\text{total}}^2 R_{\text{total}}$$
$$= (0.2 \text{ A})^2 (60 \text{ }\Omega)$$
$$= 2.4 \text{ W}$$

Answer is B.

(f)

$$P_{R_x} = I_{\text{total}}^2 R_x$$
$$P_{R_{10 \text{ }\Omega}} = (0.2 \text{ A})^2 (10 \text{ }\Omega) = 0.4 \text{ W}$$
$$P_{R_{20 \text{ }\Omega}} = (0.2 \text{ A})^2 (20 \text{ }\Omega) = 0.8 \text{ W}$$
$$P_{R_{30 \text{ }\Omega}} = (0.2 \text{ A})^2 (30 \text{ }\Omega) = 1.2 \text{ W}$$

Or,

$$P_{R_x} = \frac{V_{R_x}^2}{R_x}$$

$$P_{R_{10 \text{ }\Omega}} = \frac{(2 \text{ V})^2}{10 \text{ }\Omega} = 0.4 \text{ W}$$

$$P_{R_{20 \text{ }\Omega}} = \frac{(4 \text{ V})^2}{20 \text{ }\Omega} = 0.8 \text{ W}$$

$$P_{R_{30 \text{ }\Omega}} = \frac{(6 \text{ V})^2}{30 \text{ }\Omega} = 1.2 \text{ W}$$

The calculations can be checked by summing the individual resistor powers to check against the total power.

Answer is A.

2. (a) The total resistance in this parallel circuit is

$$R_{\text{total}} = \frac{R_1 R_2}{R_1 + R_2} = \frac{(10 \text{ }\Omega)(15 \text{ }\Omega)}{10 \text{ }\Omega + 15 \text{ }\Omega}$$
$$= 6 \text{ }\Omega$$

Another solution method is

$$R_{\text{total}} = \left(\frac{1}{\dfrac{1}{R_1} + \dfrac{1}{R_2}}\right) = \left(\frac{1}{\dfrac{1}{10 \text{ }\Omega} + \dfrac{1}{15 \text{ }\Omega}}\right)$$
$$= 6 \text{ }\Omega$$

This method is easy to use with the reciprocal, store, and sum keys on any scientific calculator.

Note: Option (A), $-5 \text{ }\Omega$, is the difference between the two resistors and is impossible because it is negative. All real nonsuperconductive resistances are positive. Option (B) is incorrect because it represents the total conductance. The problem asked for total resistance, which is the reciprocal of conductance. Option (D) is incorrect because 25 Ω is the sum of the two resistors, which would only be true if this was a series circuit.

Answer is C.

(b) Using Ohm's law, the current is

$$I = \frac{V}{R} = \frac{12 \text{ V}}{6 \text{ }\Omega}$$
$$= 2 \text{ A}$$

Answer is A.

(c) This is a parallel circuit, so the voltage across each resistor is the same as the source voltage: 12 V.

Note: Option (A) would be correct for a series circuit.

Answer is C.

(d) Each resistor has the full source voltage of 12 V across it. The current across each resistor is found by using Ohm's law.

$$I_{R_{10\ \Omega}} = \frac{V}{R_{10\ \Omega}} = \frac{12\ \text{V}}{10\ \Omega}$$
$$= 1.2\ \text{A}$$
$$I_{R_{15\ \Omega}} = \frac{V}{R_{15\ \Omega}} = \frac{12\ \text{V}}{15\ \Omega}$$
$$= 0.8\ \text{A}$$

Notice that the sum of the two resistor currents is the total current.

$$0.8\ \text{A} + 1.2\ \text{A} = 2.0\ \text{A}$$

This is KCL.

Answer is C.

(e) The total power is

$$P_{\text{total}} = V I_{\text{total}} = (12\ \text{V})(2\ \text{A})$$
$$= 24\ \text{W}$$

Answer is D.

(f) There are three ways to find power dissipated in a resistor. The first way is

$$P = \frac{V^2}{R}$$
$$P_{R_{10\ \Omega}} = \frac{(12\ \text{V})^2}{10\ \Omega}$$
$$= 14.4\ \text{W} \quad (14\ \text{W})$$
$$P_{R_{15\ \Omega}} = \frac{(12\ \text{V})^2}{15\ \Omega}$$
$$= 9.6\ \text{W}$$
$$P_{R_{\text{total}}} = P_{R_{10\ \Omega}} + P_{R_{15\ \Omega}}$$
$$= 14.4\ \text{W} + 9.6\ \text{W}$$
$$= 24\ \text{W}$$

The second way is

$$P = I^2 R$$
$$P_{R_{10\ \Omega}} = (1.2\ \text{A})^2 (10\ \Omega)$$
$$= 14.4\ \text{W} \quad (14\ \text{W})$$

$$P_{R_{15\ \Omega}} = (0.8\ \text{A})^2 (15\ \Omega)$$
$$= 9.6\ \text{W}$$
$$P_{R_{\text{total}}} = P_{R_{10\ \Omega}} + P_{R_{15\ \Omega}}$$
$$= 14.4\ \text{W} + 9.6\ \text{W}$$
$$= 24\ \text{W}$$

The third way is

$$P = VI$$
$$P_{R_{10\ \Omega}} = (12\ \text{V})(1.2\ \text{A})$$
$$= 14.4\ \text{W} \quad (14\ \text{W})$$
$$P_{R_{15\ \Omega}} = (12\ \text{V})(0.8\ \text{A})$$
$$= 9.6\ \text{W}$$
$$P_{R_{\text{total}}} = P_{R_{10\ \Omega}} + P_{R_{15\ \Omega}}$$
$$= 14.4\ \text{W} + 9.6\ \text{W}$$
$$= 24\ \text{W}$$

Answer is B.

3. (a) Resistance for more than two parallel resistors is best found by using the following equation.

$$R_{\text{total}} = \frac{1}{\dfrac{1}{R_1} + \dfrac{1}{R_2} + \dfrac{1}{R_3}}$$
$$= \frac{1}{\dfrac{1}{12\ \Omega} + \dfrac{1}{15\ \Omega} + \dfrac{1}{20\ \Omega}}$$
$$= 5.0\ \Omega$$

Answer is A.

(b) The total current is found through Ohm's law.

$$I_{\text{total}} = \frac{V}{R_{\text{parallel}}} = \frac{30\ \text{V}}{5\ \Omega}$$
$$= 6\ \text{A}$$

Answer is C.

(c) This is a parallel circuit, so the voltage across each resistor is the same as the source voltage: 30 V.

Answer is D.

(d) Each resistor has the full source voltage of 30 V across it. The current across each resistor is found by using Ohm's law.

$$I = \frac{V}{R}$$
$$I_{10\ \Omega} = \frac{30\ \text{V}}{12\ \Omega} = 2.5\ \text{A}$$
$$I_{15\ \Omega} = \frac{30\ \text{V}}{15\ \Omega} = 2.0\ \text{A}$$
$$I_{20\ \Omega} = \frac{30\ \text{V}}{20\ \Omega} = 1.5\ \text{A}$$

Notice that the sum of the two currents is the total current.

$$2.5 \text{ A} + 2.0 \text{ A} + 1.5 \text{ A} = 6.0 \text{ A}$$

This is KCL.

Answer is B.

(e)
$$\begin{aligned} P_{\text{total}} &= V I_{\text{total}} \\ &= (30 \text{ V}) (6 \text{ A}) \\ &= 180 \text{ W} \end{aligned}$$

Answer is D.

(f) There are three ways to find power dissipated in a resistor. The first method is

$$P = \frac{V^2}{R}$$

$$P_{12 \ \Omega} = \frac{(30 \text{ V})^2}{12 \ \Omega} = 75 \text{ W}$$

$$P_{15 \ \Omega} = \frac{(30 \text{ V})^2}{15 \ \Omega} = 60 \text{ W}$$

$$P_{20 \ \Omega} = \frac{(30 \text{ V})^2}{20 \ \Omega} = 45 \text{ W}$$

$$\begin{aligned} P_{\text{total}} &= P_{12 \ \Omega} + P_{15 \ \Omega} + P_{20 \ \Omega} \\ &= 75 \text{ W} + 60 \text{ W} + 45 \text{ W} \\ &= 180 \text{ W} \end{aligned}$$

Another method is

$$P = I^2 R$$

$$\begin{aligned} P_{12 \ \Omega} &= (2.5 \text{ A})^2 (12 \ \Omega) \\ &= 75 \text{ W} \end{aligned}$$

$$\begin{aligned} P_{15 \ \Omega} &= (2.0 \text{ A})^2 (15 \ \Omega) \\ &= 60 \text{ W} \end{aligned}$$

$$\begin{aligned} P_{20 \ \Omega} &= (1.5 \text{ A})^2 (20 \ \Omega) \\ &= 45 \text{ W} \end{aligned}$$

$$\begin{aligned} P_{\text{total}} &= P_{12 \ \Omega} + P_{15 \ \Omega} + P_{20 \ \Omega} \\ &= 75 \text{ W} + 60 \text{ W} + 45 \text{ W} \\ &= 180 \text{ W} \end{aligned}$$

The third method is

$$P = VI$$

$$\begin{aligned} P_{12 \ \Omega} &= (30 \text{ V}) (2.5 \text{ A}) \\ &= 75 \text{ W} \end{aligned}$$

$$\begin{aligned} P_{15 \ \Omega} &= (30 \text{ V}) (2.0 \text{ A}) \\ &= 60 \text{ W} \end{aligned}$$

$$\begin{aligned} P_{20 \ \Omega} &= (30 \text{ V}) (1.5 \text{ A}) \\ &= 45 \text{ W} \end{aligned}$$

$$\begin{aligned} P_{\text{total}} &= P_{12 \ \Omega} + P_{15 \ \Omega} + P_{20 \ \Omega} \\ &= 75 \text{ W} + 60 \text{ W} + 45 \text{ W} \\ &= 180 \text{ W} \end{aligned}$$

Answer is C.

4. These resistors are of equal value. There is an easy method of determining their total parallel value.

$$\begin{aligned} R_{\text{total}} &= \frac{R}{n} = \frac{100 \ \Omega}{4} \\ &= 25 \ \Omega \end{aligned}$$

Answer is B.

5. Capacitance adds in parallel circuits.

$$1 \ \mu\text{F} + 2 \ \mu\text{F} + 4 \ \mu\text{F} = 7.0 \ \mu\text{F}$$

Answer is C.

6.
$$\begin{aligned} W &= 0.5 \, C V^2 \\ &= (0.5) \left(100 \times 10^{-6} \text{ F}\right) (200 \text{ V})^2 \\ &= 2 \text{ J} \end{aligned}$$

Answer is B.

7.
$$\begin{aligned} C_{\text{total}} &= \frac{C_1 C_2}{C_1 + C_2} \\ &= \frac{(5 \text{ nF})(10 \text{ nF})}{5 \text{ nF} + 10 \text{ nF}} \\ &= 3.33 \text{ nF} \quad (3.3 \text{ nF}) \end{aligned}$$

Or,
$$\begin{aligned} C_{\text{total}} &= \frac{1}{\dfrac{1}{C_1} + \dfrac{1}{C_2}} \\ &= \frac{1}{\dfrac{1}{5 \text{ nF}} + \dfrac{1}{10 \text{ nF}}} \\ &= 3.33 \text{ nF} \quad (3.3 \text{ nF}) \end{aligned}$$

Answer is A.

8. The capacitors are of equal value. The easiest method to find their series value is

$$\begin{aligned} C_{\text{total}} &= \frac{C}{n} = \frac{10 \text{ pF}}{4} \\ &= 2.5 \text{ pF} \end{aligned}$$

Or,

$$C_{\text{total}} = \cfrac{1}{\cfrac{1}{C_1} + \cfrac{1}{C_2} + \cfrac{1}{C_3} + \cfrac{1}{C_4}}$$

$$= \cfrac{1}{\cfrac{1}{10 \text{ pF}} + \cfrac{1}{10 \text{ pF}} + \cfrac{1}{10 \text{ pF}} + \cfrac{1}{10 \text{ pF}}}$$

$$= 2.5 \text{ pF}$$

Answer is A.

9.

$$C_{\text{series}} = \cfrac{1}{\cfrac{1}{C_1} + \cfrac{1}{C_2} + \cfrac{1}{C_3}}$$

$$= \cfrac{1}{\cfrac{1}{1.0 \ \mu\text{F}} + \cfrac{1}{3.0 \ \mu\text{F}} + \cfrac{1}{5.0 \ \mu\text{F}}}$$

$$= 0.65 \ \mu\text{F}$$

Note: Option (B) does not make the final inversion (i.e., this is the reciprocal of option (A)). Option (C) is the parallel capacitance.

Answer is A.

10. The voltage is divided equally between the two equal capacitors, so each capacitor has 4.5 V across it.

Answer is B.

11.

$$C_{\text{total}} = \cfrac{1}{\cfrac{1}{C_1} + \cfrac{1}{C_2}} = \cfrac{1}{\cfrac{1}{10 \ \mu\text{F}} + \cfrac{1}{20 \ \mu\text{F}}}$$

$$= 6.667 \ \mu\text{F}$$

$$V_X = V_{\text{total}} \frac{C_{\text{total}}}{C_X}$$

$$V_1 = V_{\text{total}} \frac{C_{\text{total}}}{C_1}$$

$$= (9 \text{ V}) \left(\frac{6.667 \ \mu\text{F}}{10 \ \mu\text{F}} \right)$$

$$= 6 \text{ V}$$

$$V_2 = V_{\text{total}} \frac{C_{\text{total}}}{C_2}$$

$$= (9 \text{ V}) \left(\frac{6.667 \ \mu\text{F}}{20 \ \mu\text{F}} \right)$$

$$= 3 \text{ V}$$

KVL applies.

$$V_{\text{total}} = V_1 + V_2$$

$$= 6 \text{ V} + 3 \text{ V}$$

$$= 9 \text{ V}$$

Answer is C.

12.

$$\tau = RC$$

$$= \left(2.0 \times 10^3 \ \Omega \right) \left(1.0 \times 10^{-6} \text{ F} \right)$$

$$= 2.0 \times 10^{-3} \text{ s} \quad (2 \text{ ms})$$

At $t = 4.0$ ms,

$$i_C = \frac{V_{\text{in}}}{R} e^{-\frac{t}{RC}}$$

$$= \left(\frac{15 \text{ V}}{2 \text{ k}\Omega} \right) \left(e^{-\frac{4 \times 10^{-3} \text{ s}}{(2 \times 10^3 \ \Omega)(1.0 \times 10^{-6} \text{ F})}} \right)$$

$$= (7.5 \text{ mA}) \, e^{-\frac{4.0 \text{ ms}}{2.0 \text{ ms}}}$$

$$= 1.015 \text{ mA} \quad (1.0 \text{ mA})$$

Note: Option (B) would be a correct answer at $t = 2.0$ ms $= RC$, but not at $t = 4$ ms $= 2RC$. Option (C) does not take into account the exponential decay. Option (D) is the applied voltage instead of the current.

Answer is A.

13.

$$v = \frac{It}{C}$$

$$= \frac{\left(0.5 \times 10^{-3} \text{ A} \right) \left(10 \times 10^{-3} \text{ s} \right)}{2.2 \times 10^{-6} \text{ F}}$$

$$= 2.27 \text{ V} \quad (2.3 \text{ V})$$

Note: Option (A) multiplies the current by the time, but neglects the capacitance. Option (C) resulted from an error in exponent use. Option (D) does not consider the time.

Answer is B.

14.

$$RC = \left(10 \times 10^3 \ \Omega \right) \left(1 \times 10^{-6} \text{ F} \right)$$

$$= 10 \times 10^{-3} \text{ s} \quad (10 \text{ ms})$$

$$V_R = V_{\text{in}} e^{-\frac{t}{RC}}$$

$$= (10 \text{ V}) \, e^{-\frac{t}{10 \text{ ms}}}$$

At $t = 15$ ms,

$$V_R = (10 \text{ V}) \, e^{-\frac{15 \text{ ms}}{10 \text{ ms}}}$$

$$= (10 \text{ V}) \, e^{-1.5}$$

$$= 2.23 \text{ V} \quad (2.2 \text{ V})$$

Note: Option (B) is incorrect because it does not use the RC value (i.e., $v = (10 \text{ V})e^{-0.015\ s} = 98.5$ V). Option (C) does not use the time value (i.e., $v = (10 \text{ V})e^{-0.01} = 9.9$ V). Option (D) uses a positive exponent, giving a resistance voltage greater than the input voltage, which is an impossibility.

Answer is A.

15.
$$e = -N\frac{\Delta\Phi}{\Delta t}$$
$$= (-50 \text{ turns})\left(\frac{1.2 \text{ Wb}}{1.0 \text{ s}}\right)$$
$$= -60 \text{ V}$$

Note: Option (D) is incorrect because the minus sign in the equation was neglected.

Answer is A.

16.
$$e = -N\frac{\Delta\Phi}{\Delta t}$$
$$= -(70 \text{ turns})\left(\frac{-4.5 \times 10^{-3} \text{ Wb}}{30 \text{ ms}}\right)$$
$$= 10.5 \text{ V} \quad (11 \text{ V})$$

Note: Option (A) is incorrect because the minus sign was neglected.

Answer is C.

17. The total inductance is the sum of the values of the three inductors in series.

$$10 \text{ mH} + 20 \text{ mH} + 5 \text{ mH} = 35 \text{ mH} \quad (40 \text{ mH})$$

Note: Option (A) is incorrect because it is the total parallel inductance, rather than the total series inductance.

Answer is C.

18.
$$W = \tfrac{1}{2}LI^2$$
$$= \frac{(0.22 \text{ H})(0.2 \text{ A})^2}{2}$$
$$= 0.0044 \text{ J}$$

Note: Option (B) does not multiply by $\frac{1}{2}$ as required by the equation. Option (C) does not square the current. Option (D) fails because of an incorrect use of exponents and SI prefixes (i.e., not recognizing that milli is 10^{-3}).

Answer is A.

19.
$$\tau = \frac{L}{R} = \frac{360 \text{ }\mu\text{H}}{12 \text{ }\Omega}$$
$$= 30 \text{ }\mu\text{s}$$

Note: Options (A) and (D) result from an improper use of the equation. Option (C) is incorrect because of an incorrect use of exponents and prefixes.

Answer is B.

20.
$$V_L = V_1 e^{-\frac{t}{\tau}}$$
$$= (24 \text{ V})\, e^{-\frac{t}{30 \text{ }\mu\text{s}}}$$

At $t = 20$ ms,

$$V_L = (24 \text{ V})\, e^{-\frac{20 \text{ }\mu\text{s}}{30 \text{ }\mu\text{s}}}$$
$$= (24 \text{ V})\, e^{-0.6666}$$
$$= (24 \text{ V})\,(0.5134)$$
$$= 12.322 \text{ V} \quad (12 \text{ V})$$

Note: Option (A) would be true at $t = \infty$ ms. Option (C) would be true at $t = 0$ ms. Option (D) results from using a positive exponent for e.

Answer is B.

21. The current is decreasing (negative).

$$e = -L\frac{\Delta I}{\Delta t}$$
$$= (-0.4 \text{ H})\left(\frac{0.1 \text{ A} - 0.5 \text{ A}}{10 \text{ ms}}\right)\left(1 \times 10^3 \text{ }\frac{\text{ms}}{\text{s}}\right)$$
$$= 16 \text{ V}$$

Note: Not taking the negative current into account gives option (A).

Answer is C.

22.
$$R = \frac{\rho l}{A}$$
$$= \frac{(115 \times 10^{-8} \text{ }\Omega\cdot\text{m})(4000 \text{ m})}{\pi(0.0005 \text{ m})^2}$$
$$= 5857 \text{ }\Omega \quad (5900 \text{ }\Omega)$$

Note: For option (A), the kilometer exponent was left out. Option (B) is incorrect because diameter is mistaken for radius. Option (D) is incorrect because π was left out.

Answer is C.

23.
$$R = \frac{\rho l}{A}$$
$$= \frac{\left(10.37 \ \frac{\text{CM·}\Omega}{\text{ft}}\right)(10^4 \ \text{ft})}{642.40 \ \text{CM}}$$
$$= 161.4 \ \Omega \quad (160 \ \Omega)$$

Note: Option (A) is incorrect because the CM area was squared. For option (B), π was incorrectly inserted into the equation. Option (C) is incorrect because the square root of the area was used.

Answer is D.

24.
$$R = R_o \left(1 + \alpha \left(T - T_o\right)\right)$$
$$= (10 \ \Omega) \left(\begin{array}{l} 1 + \left(3.93 \times 10^{-3} \ \frac{1}{°\text{C}}\right) \\ \times \ (120°\text{C} - 20°\text{C}) \end{array} \right)$$
$$= 13.93 \ \Omega \quad (14 \ \Omega)$$

Note: Options (A) and (B) used wrong equations. The temperature difference in options (B) and (D) is incorrect: the 20°C was not subtracted from 120°C.

Answer is C.

25.
$$F = k \frac{C_1 C_2}{d^2}$$
$$= \left(9.0 \times 10^9 \ \frac{\text{N·m}^2}{\text{C}^2}\right)$$
$$\times \left(\frac{\left(30 \times 10^{-6} \ \text{C}\right)\left(40 \times 10^{-6} \ \text{C}\right)}{(0.12 \ \text{m})^2}\right)$$
$$= 750 \ \text{N repulsive}$$

Note: Option (A) does not use the constant multiplier, $k = 9.0 \times 10^9$. Option (B), 90 N repulsive, is incorrect because the numerator is divided by 0.12 m, rather than by $(0.12 \ \text{m})^2$. Option (D) is incorrect because the force between two like charges must be repulsive, not attractive.

Answer is C.

26. The flux density in the rod is
$$B = \frac{\Phi}{A} = \frac{22 \times 10^{-6} \ \text{Wb}}{\pi \left(\dfrac{0.01 \ \text{m}}{2}\right)^2}$$
$$= 0.28 \ \text{T}$$

Note: Option (A) is incorrect because the diameter was not squared. Option (B) does not divide the diameter by two. Option (D) does not use π in the denominator. This would be correct for a square rod, but not a circular rod.

Answer is C.

27.
$$B = \left(2 \times 10^{-7} \ \frac{\text{Wb}}{\text{m·A}}\right)\frac{I}{r}$$
$$= \left(2 \times 10^{-7} \ \frac{\text{Wb}}{\text{mA}}\right)\left(\frac{0.12 \ \text{A}}{0.05 \ \text{m}}\right)$$
$$= 4.8 \times 10^{-7} \ \text{T}$$

Note: Option (B) is incorrect because the constant was not used. For option (C), the constant's exponent was not used. For option (D), the distance was squared.

Answer is A.

2 Alternating-Current Circuits

Nomenclature

B	susceptance	S
BW	bandwidth	Hz
C	capacitance	F
emf	electromotive force	V
f	frequency	Hz
G	conductance	S
$i(t)$	alternating current	A
L	inductance	H
N	number of turns	–
P	real power	W
Q	quality factor	–
Q	reactive power	VAR
R	resistance	Ω
S	apparent power	V·A
T	period	s
$v(t)$	alternating voltage	V
V	voltage	V
X	reactance	Ω
Y	admittance	S
Z	impedance	Ω

Symbols

θ	phase angle, phase shift	degree, rad
ω	angular frequency	rad/s
η	efficiency	–

Subscripts

0	at resonance
AC	alternating current
avg	average
C	capacitive
DC	direct current
L	inductive
max	maximum
min	minimum
P	peak
P	primary coil
$P+$	positive peak
$P-$	negative peak
PP	peak to peak
R	resistive
S	secondary coil

ALTERNATING CURRENT

Alternating current (AC) is current that changes polarity and thus direction. AC can be pulsed, rectangular, sawtooth, triangular, sinusoidal, or any real waveform. Three possible AC waveforms are shown in Fig. 2.1. This is how the waveforms would appear on an oscilloscope.

Figure 2.1 AC Waveforms

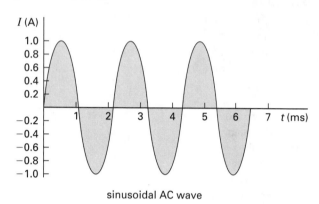

sinusoidal AC wave

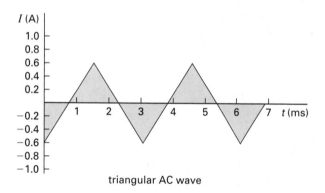

triangular AC wave

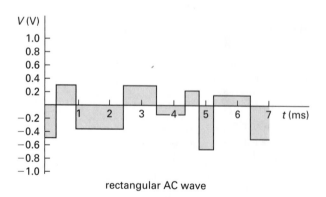

rectangular AC wave

Figure 2.2 shows the parameters of a representative AC waveform. V_{P+} is the peak positive-going voltage, or maximum voltage. V_{P-} is the peak negative-going voltage, or minimum voltage. V_{PP} is the peak-to-peak voltage. The peak-to-peak voltage is the difference between the maximum voltage and the minimum voltage.

$$V_{PP} = V_{P+} - V_{P-}$$

The period, T, is the time the waveform takes before it starts to repeat itself.

Figure 2.2 Waveform Parameters

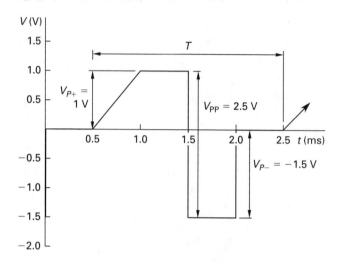

Note the beginning of the trace is not necessarily the starting point to begin the period measurement. The period end is the point beyond which the waveform repeats its cycle. An oscilloscope can be made to trigger (start the trace) at any point of the waveform.

The frequency, f, is how frequently or how often a waveform repeats itself in 1 s. It is the reciprocal of the period in seconds, T. The unit of frequency is the hertz (Hz).

$$f = \frac{1}{T} \qquad 2.2$$

Example 2.1

Find the peak voltage, peak negative voltage, peak-to-peak voltage, and period for the waveform in Fig. 2.2.

Solution

The positive peak voltage, V_{P+}, is 1.0 V. This is the same as the maximum voltage.

The negative peak voltage, V_{P-}, is −1.5 V. This is the same as the minimum voltage.

The peak-to-peak voltage, V_{PP}, is

$$
\begin{aligned}
V_{PP} &= V_{P+} - V_{P-} \\
&= 1.0 \text{ V} - (-1.5 \text{ V}) \\
&= 2.5 \text{ V}
\end{aligned}
$$

The period is

$$T = 2.5 \text{ ms} - 0.5 \text{ ms}$$
$$= 2.0 \text{ ms}$$

Note the period does not necessarily begin at the start of the trace.

Example 2.2

Find the frequency of the waveform in Fig. 2.2.

Solution

The period is one cycle of the waveform. The waveform cycle in the figure begins at 0.5 ms and ends at a time of 2.5 ms. The period is

$$T = 2.5 \text{ ms} - 0.5 \text{ ms}$$
$$= 2.0 \text{ ms}$$

The frequency is

$$f = \frac{1}{T} = \frac{1}{2.0 \times 10^{-3} \text{ s}}$$
$$= 0.5 \times 10^3 \text{ Hz} \quad (500 \text{ Hz})$$

Working in the given units of milliseconds, the frequency is

$$f = \frac{1}{T} = \frac{1}{(2.0 \text{ ms}) \left(0.001 \ \frac{\text{s}}{\text{ms}} \right)}$$
$$= 500 \text{ Hz}$$

NON-AC WAVEFORMS

AC is current that changes direction. Figure 2.3 shows changing waveforms that are not AC because their currents or voltages do not change direction (i.e., the waves do not cross the 0 A axis).

SINUSOIDAL WAVEFORMS

This chapter will mostly study sinusoidal waveforms (sine waves). This is not a serious limitation because any real waveform can be broken down into a combination of sine wave harmonics. Waveforms from the commercial power grids and single-frequency (pitch) audio tones are examples of sinusoidal waveforms.

Figure 2.4 is a sinusoidal waveform showing the various parameters: time, voltage as a function of time, frequency in hertz, phase shift in degrees, the maximum or peak positive voltage, the minimum or peak negative voltage, and the peak-to-peak voltage. The period is the time during which the waveform goes through one cycle, or 360°, or 2π radians.

Figure 2.3 Non-AC Waveforms

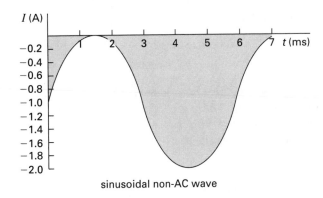

sinusoidal non-AC wave

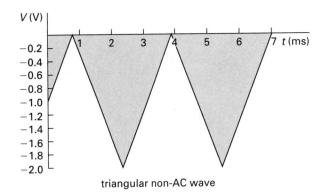

triangular non-AC wave

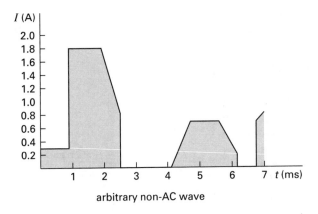

arbitrary non-AC wave

Sinusoidal waveforms are represented in three ways: trigonometric form, vector (polar) form, and rectangular form. The rectangular form uses complex notation (real and imaginary). Remember maximum voltage is the same as peak positive voltage. t is the time and θ is the phase shift in degrees.

The voltage as a function of time, $v(t)$, in trigonometric notation is

$$v(t) = V_{P+} \sin(2\pi f t + \theta)$$
$$= V_P \sin(2\pi f t + \theta)$$
$$= V_{\max} \sin(2\pi f t + \theta) \qquad 2.3$$

Maximum current is the same as peak positive current. The current as a function of time, $i(t)$, in trigonometric notation is

$$i(t) = I_P \sin (2\pi f t + \theta)$$
$$= I_{\max} \sin (2\pi f t + \theta) \qquad 2.4$$

These trigonometric notations can use the natural radian frequency. There are 2π radians per cycle ($360°$) or period, so $2\pi f = \omega$ rad/s. f is in Hz (1/s). The trigonometric form using the natural frequency is

$$v(t) = V_{P+} \sin (\omega t + \theta) = V_P \sin(\omega t + \theta)$$
$$= V_{\max} \sin (\omega t + \theta)$$
$$= V_{\max} \sin (2\pi t + \theta) \qquad 2.5$$
$$i(t) = I_{P+} \sin (\omega t + \theta) = I_P \sin$$
$$= I_{\max} (\omega t + \theta)$$
$$= I_{\max} \sin(2\pi t + \theta) \qquad 2.6$$

The rectangular form is

$$v = V_{\text{real}} + V_{\text{imaginary}} = V_P \angle \theta \qquad 2.7$$

The vector form is

$$v = V_{P+} \angle \theta = V_P \angle \theta \qquad 2.8$$

The polar form is almost the same as the vector form.

$$v = V_{P+} \angle \theta \qquad 2.9$$

Figure 2.4 Sinusoidal Waveform Parameters

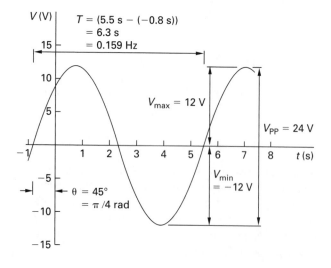

In Fig. 2.4, the sine wave period occurs from -0.8 s to 5.5 s. The period is

$$T = 5.5 \text{ s} - (-0.8 \text{ s})$$
$$= 6.3 \text{ s}$$

The frequency is

$$f = \frac{1}{T} = \frac{1}{6.3 \text{ s}}$$
$$= 0.16 \text{ Hz}$$

When inductors and capacitors are used in a circuit, the current can be out of phase with the applied voltage. In Fig. 2.4, the phase shift is the angular time difference between the time at 0 s and when the waveform crosses 0 V. The waveform crosses 0 V at -0.8 s. The phase shift is

$$\theta = \frac{\Delta t}{T} (360°)$$
$$= \left(\frac{0.8 \text{ s} - 0 \text{ s}}{T} \right) (360°)$$
$$= \left(\frac{0.8 \text{ s}}{6.3 \text{ s}} \right) (360°)$$
$$= 45°$$

The voltage as a function of time is

$$v(t) = V_{\max} \sin (2\pi f t + \theta)$$
$$= 12 \text{ V} \sin(2\pi (0.0159 \text{ Hz})t + \theta)$$

$v(t)$ is leading by $45°$.

This equation can be put into the natural frequency or radian frequency. The angular period is the time needed for the waveform to pass through one cycle, $360°$ or 2π radians per cycle. The angular frequency, ω, is the reciprocal of the angular period.

$$\omega = \frac{1 \text{ cycle}}{T}$$
$$= \frac{2\pi \text{ radians}}{T}$$
$$= 2\pi f \qquad 2.10$$

The angular frequency in Fig. 2.4 is

$$\omega = 2\pi f$$
$$= 2\pi (0.159 \text{ Hz})$$
$$= 1.0 \text{ rad/s}$$
$$v(t) = 12 \text{ V} \sin(1.0t + 45°)$$
$$= 12 \text{ V} \sin(t + 45°)$$

The frequency is constant in many applications. For example, the frequency in power systems is a constant 60 Hz in the Americas and 50 Hz in Europe and Japan. Lower frequencies are more efficient for AC motors. Higher frequencies do not need as much iron in magnets, motors, and transformers, making the equipment lighter. For instance, 400 Hz power systems are used in aircraft to save weight. However, 400 Hz motors have an extremely annoying and constant shriek requiring hearing protection or excellent sound insulation.

This equation has been introduced. Note that the frequency is given in radians per second and the phase shift in degrees. This mixing of units is technically improper, but this is the standard notation.

$$v(t) = V_{\max} \sin((2\pi \text{ Hz})t + \theta)$$
$$= V_{\max} \sin((\omega_{\text{rad/s}})t + \theta)$$

If the AC frequency is constant, a vector form of the voltage or current can be used. This vector equation does not have frequency information, and a sine wave is assumed. This makes the polar equation more compact than the time equation. The waveform of Fig. 2.4 can be described in vector (or polar) form.

$$v = V_{\max} \angle \theta$$
$$= 12 \text{ V} \angle 45°$$
$$= V_P \angle \theta$$

V_P and $V_{\max}$ are equivalent and are used interchangeably. A polar or vector plot of this equation is shown in Fig. 2.5.

Figure 2.5 *Polar (Vector) Plot*

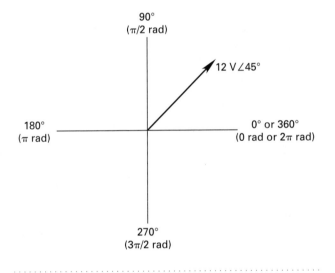

TIME AND PHASE RELATIONSHIPS

The capacitors and inductors in DC circuits store energy and release the stored energy. This store-release process takes a finite time. AC waveforms take time to cycle through their period. Hence time becomes a consideration in studies of AC circuits, and inductors and capacitors.

An output waveform can be delayed or advanced relative to the input signal source by combinations of inductor, capacitor and/or resistor. Figure 2.6 shows various component configurations for advancing or delaying the output waveform relative to the input waveform.

Figure 2.6 *Configurations for Phase Advance and Phase Delay*

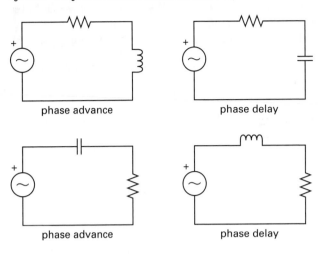

Phase shift is the angular difference between the input signal and the output signal. It is expressed as degrees, rather than radians or time. A 1000 Hz, 1.0 V peak signal with a −60° phase shift as shown in Fig. 2.7 is described by

$$v_2(t) = V_{\max} \sin(\omega \pm \theta)$$
$$= 1.0 \text{ V} \sin(2\pi (1000 \text{ Hz}) t \pm \theta)$$
$$= 1.0 \text{ V} \sin(2\pi (1000t^{-1}) t - 60°)$$
$$= 1.0 \text{ V} \sin(6283t - 60°)$$
$$= 1.0 \text{ V} \angle -60°$$

Figure 2.7 *Waveforms Separated by 60° Phase Shift*

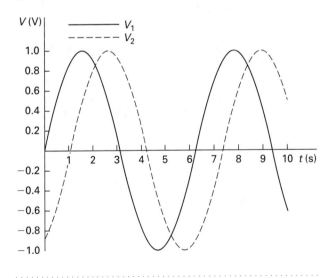

Note that frequency is in radians ($\omega = 2\pi f$) and the phase shift is given in degrees. This can be a pitfall if an engineer's calculator is not set properly. An EET professor once had his calculator settings changed by his students during a break, and he became frustrated at his calculations coming out incorrectly. The mischievous students could barely contain their glee.

Figure 2.8 shows the two waveforms in polar coordinates.

Figure 2.8 Polar (Vector) Plot Waveforms Separated by 60° Phase Shift

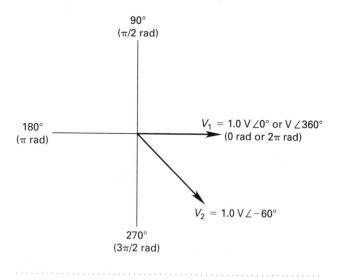

Figure 2.9 Full-Wave Waveform

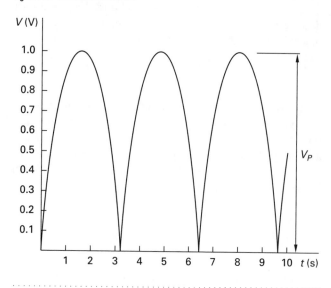

Figure 2.10 Half-Wave Sine

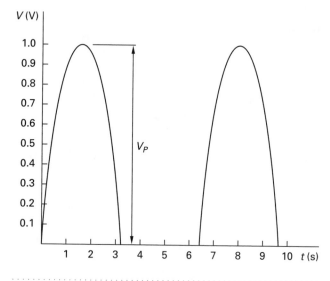

AVERAGE VALUE

The average voltage value of a periodic waveform is

$$V_{\text{avg}} = \frac{V_{\text{max}}}{T} \int_0^T v(t)\,dt \qquad 2.11$$

A full-wave, periodic waveform is shown in Fig. 2.9. This is the absolute value of a 1.0 V sine wave. It does not have a negative portion. This waveform can be generated by a full-wave rectifier circuit which converts AC to DC. This is discussed in Ch. 4 in the section on diodes. The waveform changes but never becomes negative, so it is DC.

Figure 2.10 is a half-wave sine. The average value of a half-wave, rectified sine wave (only the positive portion of the cycle) is

$$V_{\text{avg}} = \frac{V_{\text{max}}}{\pi} \qquad 2.12$$

Example 2.3

What is the average value of the full-wave, rectified sine wave (absolute value of a sine wave) in Fig. 2.9?

Solution

$$V_{\text{avg}} = \frac{2V_{\text{max}}}{\pi} = \frac{(2)\,(1\text{ V})}{\pi}$$
$$= 0.6366 \text{ V} \quad (0.64 \text{ V})$$

Example 2.4

What is the average value of the waveform in Fig. 2.10?

Solution

The average value of the waveform is

$$V_{\text{avg}} = \frac{V_{\text{max}}}{\pi} = \frac{1 \text{ V}}{\pi}$$
$$= 0.3183 \text{ V} \quad (0.318 \text{ V})$$

Average Value of a Sine Wave

A sine wave has equal positive portions and negative portions. A voltage waveform with equal parts above and below the time axis will integrate to an average value of 0 V. The average voltage value of a sine wave is

$$
\begin{aligned}
V_{\text{avg}} &= \frac{V_{\max}}{T} \int_0^T \sin \omega t \,(dt) \\
&= \left(\frac{V_{\max}}{T}\right)\left(\frac{1}{\omega}\right)(-\cos \omega)\Big|_0^T \\
&= \left(\frac{V_{\max}}{T}\right)\left(\frac{1}{\omega}\right)(-1+1) \\
&= 0 \text{ V} \qquad\qquad\qquad 2.13
\end{aligned}
$$

The power grid generates sine waves. Clearly the voltage is not 0 V for these sine waves because grid users can run motors, receive power, and obtain heat. Obviously average value calculations are not useful for sine waves. Some other method of calculating effective value must be used, and that method is the root-mean-square (rms) process.

ROOT-MEAN-SQUARE VALUE

Root-mean-square refers to how the effective value (also known as the rms value) is calculated. The process can be used with any real waveform, and is valid for either voltage or current. The process steps are in reverse order to the process name.

1. Square the peak values, thus eliminating negative values.

2. Find the mean or average value.

3. Find the square root (reverse step 1, but do not restore negative values).

Example 2.5

Find the rms value of the waveform in the illustration.

Solution
Square the waveform peak values, eliminating the negative values. For area A_1,

$$
V_1 = (2 \text{ V})^2 = 4 \text{ V}^2
$$

For area A_2,

$$
V_2 = (-1 \text{ V})^2 = 1 \text{ V}^2
$$

For area A_3,

$$
V_3 = (-2 \text{ V})^2 = 4 \text{ V}^2
$$

Illustration for Example 2.5

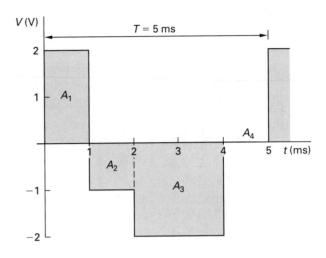

For area A_4,

$$
V_4 = (0 \text{ V})^2 = 0 \text{ V}^2
$$

Find the area of the squared values.

$$
\begin{aligned}
A_1 &= (2 \text{ V})^2 \\
&= (4 \text{ V}^2)(1 \text{ ms}) \\
&= 4 \text{ V}^2 \cdot \text{ms}
\end{aligned}
$$

$$
\begin{aligned}
A_2 &= (-1 \text{ V})^2 \\
&= (1 \text{ V}^2)(1 \text{ ms}) \\
&= 1 \text{ V}^2 \cdot \text{ms}
\end{aligned}
$$

$$
\begin{aligned}
A_3 &= (-2 \text{ V})^2 \\
&= (4 \text{ V}^2)(2 \text{ ms}) \\
&= 8 \text{ V}^2 \cdot \text{ms}
\end{aligned}
$$

$$
\begin{aligned}
A_4 &= (0 \text{ V})^2 \\
&= (0 \text{ V}^2)(1 \text{ ms}) \\
&= 0 \text{ V}^2 \cdot \text{ms}
\end{aligned}
$$

The total area is

$$
\begin{aligned}
A_{\text{total}} &= 4 \text{ V}^2 \cdot \text{ms} + 1 \text{ V}^2 \cdot \text{ms} + 8 \text{ V}^2 \cdot \text{ms} + 0 \text{ V}^2 \cdot \text{ms} \\
&= 13 \text{ V}^2 \cdot \text{ms}
\end{aligned}
$$

Divide the total area by the period to get the mean voltage squared.

$$
\begin{aligned}
V^2 &= \frac{A_{\text{total}}}{T} = \frac{13 \text{ V}^2 \cdot \text{ms}}{5 \text{ ms}} \\
&= 2.6 \text{ V}^2
\end{aligned}
$$

Reverse the first step by finding the square root, but do not restore the negative values.

$$
\begin{aligned}
V_{\text{rms}} &= \sqrt{2.6 \text{ V}^2} \\
&= 1.612 \text{ V} \quad (1.6 \text{ V})
\end{aligned}
$$

Example 2.6

Find the average current value and the rms value of the waveform in the illustration.

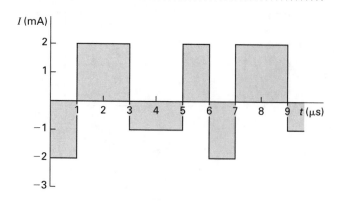

Solution

The waveform begins to repeat after 6 μs, so the period is 6 μs. It is not 9 μs, which is an easy mistake to make.

The area value from 0 μs to 1 μs is

$$A = (-2 \text{ mA}) (1 \text{ } \mu\text{s})$$
$$= -2 \text{ mA·}\mu\text{s}$$

The area value from 1 μs to 3 μs is

$$A = (2 \text{ mA}) (2 \text{ } \mu\text{s})$$
$$= 4 \text{ mA·}\mu\text{s}$$

The area value from 3 μs to 5 μs is

$$A = (-1 \text{ mA}) (2 \text{ } \mu\text{s})$$
$$= -2 \text{ mA·}\mu\text{s}$$

The area value from 5 μs to 6 μs is

$$A = (2 \text{ mA}) (1 \text{ } \mu\text{s})$$
$$= 2 \text{ mA·}\mu\text{s}$$

The sum of the area values is

$$A_{\text{total}} = -2 \text{ mA·}\mu\text{s} + 4 \text{ mA·}\mu\text{s}$$
$$- 2 \text{ mA·}\mu\text{s} + 2 \text{ mA·}\mu\text{s}$$
$$= 2 \text{ mA·}\mu\text{s}$$

The average value for the period is

$$I_{\text{avg}} = \frac{A_{\text{total}}}{T} = \frac{2 \text{ mA·}\mu\text{s}}{6 \text{ } \mu\text{s}}$$
$$= 0.33 \text{ mA}$$

To compute the rms value, first square the waveform values. The square of the current for 0 μs to 1 μs is 4 mA^2, the square of the current for 1 μs to 3 μs is 4 mA^2, the square of the current for 3 μs to 5 μs is 1 mA^2, and the square of the current for 5 μs to 6 μs is 4 mA^2.

Find the area of the squared values. The area from 0 μs to 1 μs is

$$A = (4 \text{ mA}^2) (1 \text{ } \mu\text{s})$$
$$= 4 \text{ mA}^2\text{·}\mu\text{s}$$

The area value from 1 μs to 3 μs is

$$A = (4 \text{ mA}^2) (2 \text{ } \mu\text{s})$$
$$= 8 \text{ mA}^2\text{·}\mu\text{s}$$

The area value from 3 μs to 5 μs is

$$A = (1 \text{ mA}^2) (2 \text{ } \mu\text{s})$$
$$= 2 \text{ mA}^2\text{·}\mu\text{s}$$

The area value from 5 μs to 6 μs is

$$A = (4 \text{ mA}^2) (1 \text{ } \mu\text{s})$$
$$= 4 \text{ mA}^2\text{·}\mu\text{s}$$

The total of the area values is

$$4 \text{ mA}^2\text{·}\mu\text{s} + 8 \text{ mA}^2\text{·}\mu\text{s} + 2 \text{ mA}^2\text{·}\mu\text{s} + 4 \text{ mA}^2\text{·}\mu\text{s}$$
$$= 18 \text{ mA}^2\text{·}\mu\text{s}$$

Divide the total area by the period.

$$I_{\text{mean}}^2 = \frac{A_{\text{total}}^2}{T} = \frac{18 \text{ mA}^2\text{·}\mu\text{s}}{6\mu\text{s}}$$
$$= 3 \text{ mA}^2$$

Take the square root of the mean value squared to get the effective value.

$$I_{\text{rms}} = \sqrt{I_{\text{mean}}^2} = \sqrt{(3 \text{ mA})^2}$$
$$= 1.732 \text{ mA} \quad (1.7 \text{ mA})$$

Root-Mean-Square Value of a Sine Wave

The rms values of a sine wave are

$$V_{\text{rms}} = \frac{V_{\text{max}}}{\sqrt{2}} = \frac{\sqrt{2}V_{\text{max}}}{2}$$
$$= 0.7071V_{\text{max}} \qquad 2.12$$

$$I_{\text{rms}} = \frac{I_{\text{max}}}{\sqrt{2}} = \frac{\sqrt{2}I_{\text{max}}}{2}$$
$$= 0.7071I_{\text{max}} \qquad 2.13$$

The value for peak positive voltage can be used interchangeably with maximum voltage.

Note these equations only work for sinusoidal waveforms. Students often use this formula for any waveform—triangular, square, rectangular, or other forms—and in so doing, get the answer wrong. As always, think problems through. Do not blindly throw a formula at a problem and blindly trust your calculator. Review the approach and the value calculated, and ask, "Does this make sense?"

Example 2.7

A sine wave has a peak voltage of 4.0 V. Find the rms voltage.

Solution

$$V_{rms} = \frac{V_{max}}{\sqrt{2}} = \frac{4 \text{ V}}{\sqrt{2}}$$

$$= \frac{\sqrt{2}\,(4 \text{ V})}{2}$$

$$= 2.828 \text{ V} \quad (2.8 \text{ V})$$

Example 2.8

A sine wave has an rms value for current of 2.3 A. Find the peak amperage.

Solution

$$I_P = (2.3 \text{ A})\sqrt{2}$$

$$= 3.253 \text{ A} \quad (3.25 \text{ A})$$

COMPLEX NUMBERS

Complex numbers and complex algebra are useful for representing AC signals with phase shifts. A complex number combines real and imaginary numbers. An imaginary number contains $\sqrt{-1}$, which does not exist. There are other analysis methods, but they are much more difficult than using complex numbers.

Complex numbers are represented by the following equation, in which c is the complex number, a is the real portion, and b is the imaginary portion. Electrical engineers prefer to use j instead of the i used by mathematicians, because in electrical engineering i represents current and could cause confusion in this context. The j form is used in all electrical texts.

$$c = a + \sqrt{-1}b$$
$$= a + jb \qquad 2.14$$

The conjugate of a complex number is

$$c^* = a - \sqrt{-1}b$$
$$= a - jb \qquad 2.15$$

The complex number and its conjugate are represented in the complex plane as shown in the following illustration. This is called the rectangular form of representing complex numbers because it uses rectangular coordinates.

Figure 2.11 Rectangular Plot of a Complex Number

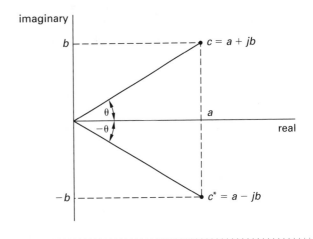

Polar or Phasor Form of Complex Numbers

Complex numbers can be written in the polar or phasor form.

$$c = a \pm jb = c\angle \pm \theta \qquad 2.16$$

This equation is based on the Pythagorean theorem. Use the triangle in Fig. 2.11.

$$c = \sqrt{a^2 + b^2}$$
$$\theta = \tan^{-1}\left(\frac{b}{a}\right) = \arctan\left(\frac{b}{a}\right)$$
$$a = c\cos\theta$$
$$b = c\sin\theta$$
$$c = a + jb = c\cos\theta + jc\sin\theta$$

Therefore, in polar or phasor form,

$$c = c(\cos\theta + j\sin\theta) = c\angle\theta$$

The real part of a complex number is negative in the second and third quadrant, and standard scientific calculators have trouble with this. A way around this problem follows. If the real part of the number complex number is negative, add 180° or π radians.

For example, find the phase angles of the complex numbers $1+j$, $-1+j$, $-1-j$, and $1-j$. Plot the numbers in the complex plane.

Figure 2.12 Phasor Plot in the Complex Plane

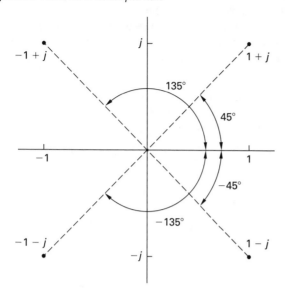

The phase angles for these complex numbers are

$$\theta_{1+j} = \tan^{-1}\left(\frac{1}{1}\right)$$
$$= 45°$$
$$\theta_{-1+j} = \tan^{-1}\left(\frac{1}{-1}\right)$$
$$= 180° + \tan^{-1}\left(\frac{-1}{1}\right)$$
$$= 180° - 45°$$
$$= 135°$$
$$\theta_{-1-j} = \tan^{-1}\left(\frac{-1}{-1}\right)$$
$$= 180° + \tan^{-1}\left(\frac{1}{1}\right)$$
$$= 180° + 45°$$
$$= 225°$$
$$\theta_{1-j} = \tan^{-1}\left(\frac{-1}{1}\right)$$
$$= -45°$$

Phase angle θ_{-1-j} can also be expressed as $225° - 360° = -135°$.

Example 2.9

A complex number is written $c = 3 + j4$. Find the polar form and draw it on the complex plane.

Solution

$$c = \sqrt{a^2 + b^2} = \sqrt{(3)^2 + (4)^2}$$
$$= \sqrt{25}$$
$$= 5$$

$$\theta = \tan^{-1}\left(\frac{b}{a}\right) = \tan^{-1}\left(\frac{4}{3}\right)$$
$$= 53.13°$$
$$c = c\angle \pm \theta$$
$$= 5\angle 53.13°$$

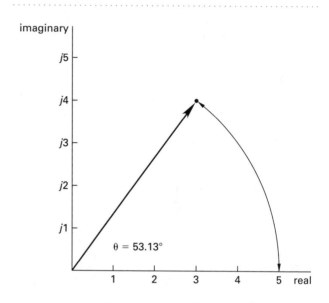

Example 2.10

Convert the polar form of the results of Ex. 2.9 into the rectangular form as a check.

Solution

$$a = c\cos\theta = 5\cos 53.13°$$
$$= 3.0$$
$$b = c\sin\theta = 5\sin 53.13°$$
$$= 4.0$$

Complex Addition and Subtraction

Arithmetic operations can use complex numbers. Real parts of complex numbers are added to or subtracted from real parts of other complex numbers; imaginary parts are added to or subtracted from imaginary parts. The two types are never mixed.

As an example, add and subtract two complex numbers, $x = 3 + j5$ and $y = 2 - j3$.

$$x + y = (3 + j5) + (2 - j3)$$
$$= (3 + 2) + j(5 - 3)$$
$$= 5 + j2$$
$$x - y = (3 + j5) - (2 - j3)$$
$$= (3 - 2) + j(5 - (-3))$$
$$= 1 + j8$$

$$y - x = (2 - j3) - (3 - j5)$$
$$= (2 - 3) + j(-3 - 5)$$
$$= -1 - j8$$

Complex Multiplication

Complex multiplication in rectangular notation proceeds much like normal multiplication. As an example, use the complex numbers $x = 3 + j5$ and $y = 2 - j3$.

$$xy = (3 + 5j)(2 - 3j)$$

$$
\begin{array}{r}
(3 + 5j) \\
\times\ (2 - 3j) \\
\hline
6 + 10j \\
+\quad -9j - 15j^2 \\
\hline
6 +\ \ j - 15j^2
\end{array}
$$

Since $j = \sqrt{-1}$ or $j^2 = -1$, the product is

$$6 + j - 15j^2 = 6 + j + 15 = 21 + j$$

Another way to multiply complex numbers is to convert them to their polar forms, and then multiply the magnitudes and add the angles.

$$x = 3 + 5j$$
$$= 5.83\angle 59.0°$$
$$y = 2 - 3j$$
$$= 3.61\angle -56.3°$$
$$xy = (5.83\angle 59.0°)(3.61\angle -56.3°)$$
$$= (5.83)(3.61)\angle(59.0° + (-56.3°))$$
$$= 21.02\angle 2.73°$$

Complex Division

The easiest manual method of dividing complex numbers is to divide the magnitudes and subtract the angles.

$$x = 3 + 5j = 5.83\angle 59.0°$$
$$y = 2 - 3j = 3.61\angle -56.3°$$
$$\frac{x}{y} = \frac{5.83}{3.61}\angle(59.0° - (-56.3°))$$
$$= 1.615\angle 115.3°$$
$$= -0.690 + j1.460$$
$$\frac{y}{x} = \frac{3.61}{5.83}\angle(-56.3° - 59.0°)$$
$$= 0.619\angle -115.3°$$
$$= -0.265 - j0.559$$

Note this result is in the third quadrant.

CAPACITIVE REACTANCE

A capacitor has an AC resistance that is more properly called *reactance*, which has the variable X_C.

When the AC source drives a capacitor, a charge is placed onto the capacitor plates, and an electric field is created between the capacitor plates. When the AC voltage source decreases, the charge in the plates goes back into the AC source (see Fig. 2.13).

When the frequency increases, the charge flow rate increases as the time for charge to flow forth and back decreases. The increased flow rate (increased current) implies less resistance (reactance). Thus, the reactance (AC resistance) decreases as frequency increases. Capacitive reactance is inversely proportional to frequency.

At no time does the current actually pass through the capacitor. The charge-discharge cycle has current flowing into and from the capacitor, giving the illusion that current (charge) is flowing through the capacitor. Even some instructors and texts describe current flowing through a capacitor, even though they know better.

Figure 2.13 Charging and Discharging a Capacitor

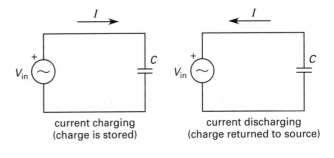

current charging
(charge is stored)

current discharging
(charge returned to source)

The current and voltage are 90° out of phase as shown in Fig. 2.14.

Figure 2.14 Current and Voltage Phase Shift

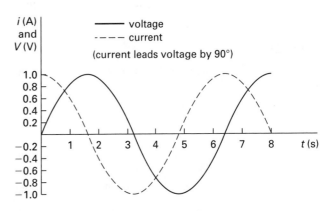

Figure 2.15 Phasor Relationship: Voltage and Current into a Capacitor

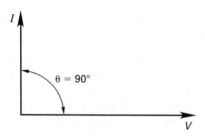

Figure 2.15 Phasor Relationship: Voltage and Current into a Capacitor

In phasor notation, the current $i = i_{\text{magnitude}} \angle 90°$ relative to the excitation voltage. The current leads the voltage by 90°. Figure 2.15 shows the phasor relationship.

Capacitive reactance magnitude, in ohms, is

$$X_C = \frac{1}{\omega C} = \frac{1}{2\pi f C} \qquad 2.17$$

Inductors also have reactance: this will be studied later in this chapter. Reactance in capacitors and inductors is similar to resistance because both have units of ohms, and follow Ohm's law and Kirchhoff's laws. The difference is that capacitors and inductors store and release energy, whereas resistors do not store energy, but only dissipate energy.

Complex algebra aids the analysis of the phase angle between the applied voltage and current. Capacitive reactance magnitude in ohms is defined as

$$|X_C| = \frac{1}{2\pi f C} = \frac{1}{\omega C} \; \Omega \qquad 2.18$$

The phase angle between the current and voltage is

$$\begin{aligned} X_C &= \frac{1}{2j\pi C} = \frac{1}{j\omega C} \\ &= -j|X_C| \\ &= |X_C| \angle {-90°} \qquad 2.19 \end{aligned}$$

Example 2.11

A 1000 Hz source drives a 10 μF capacitor. Find the reactance.

Solution

$$\begin{aligned} X_C &= \frac{1}{2\pi f C} \\ &= \frac{1}{\left(6283 \; \dfrac{\text{rad}}{\text{s}}\right)(10^{-5} \; \text{F})} \\ &= 15.9 \; \Omega \end{aligned}$$

Example 2.12

If the source frequency of Ex. 2.11 is changed from 1000 Hz to 100 Hz, what is the new reactance? Does the reactance change in inverse proportion to the frequency?

Solution

$$\begin{aligned} X_C &= \frac{1}{2\pi f C} \\ &= \frac{1}{\left(628.3 \; \dfrac{\text{rad}}{\text{s}}\right)(10^{-5} \; \text{F})} \\ &= 159 \; \Omega \end{aligned}$$

The frequency is reduced by a factor of 10; the reactance increases by a factor of 10. This shows that capacitive reactance is inversely proportional to frequency.

Example 2.13

A source frequency of 1.2 MHz drives a 100 pF capacitor. Find the reactance of the capacitor.

Solution

$$\begin{aligned} X_C &= \frac{1}{\omega C} = \frac{1}{2\pi f C} \\ &= \frac{1}{2\pi \,(1.2 \; \text{MHz})\left(10^6 \; \dfrac{\text{Hz}}{\text{MHz}}\right)} \\ &\quad \times (100 \; \text{pF})\left(10^{-12} \; \dfrac{\text{F}}{\text{pF}}\right) \\ &= 1326 \; \Omega \end{aligned}$$

Example 2.14

A 13.25 V peak source has an output frequency of 60 Hz. Find the reactance of a 10 μF capacitor and the current. What is the phase angle difference between the source voltage and current?

Solution

$$\begin{aligned} X_C &= \frac{1}{2\pi f C} \\ &= \frac{1}{2\pi \,(60 \; \text{Hz})(10 \times 10^{-6} \; \text{F})} \\ &= 265 \; \Omega \end{aligned}$$

Use Ohm's law to find the current.

$$\begin{aligned} |i_P| &= \left| \frac{v}{X_C} \right| = \frac{13.25 \; \text{V}}{265 \; \Omega} \\ &= 5.0 \times 10^{-2} \; \text{A} \quad (50 \; \text{mA}) \\ i_C &= 50 \; \text{mA} \; \angle {-90°} \end{aligned}$$

Indeed the capacitor voltage and current phase angles differ by −90°. An oscilloscope trace for the voltage and current is shown in the illustration.

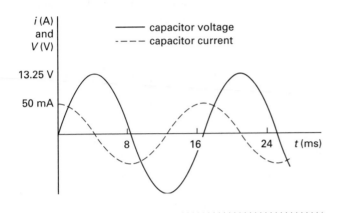

Capacitive Admittance and Susceptance

The reciprocal of reactance is susceptance. Susceptance is represented by the variable B and carries units of siemens (S).

$$|B_C| = \left|\frac{1}{X_C}\right| \qquad 2.20$$

Capacitive reactance has a phase angle of $-90°$. The complex notation is $-jX_c$. Therefore, capacitive susceptance is

$$B_C = \frac{1}{|X_C|\angle{-90°}}$$
$$= |B_C|\angle{90°}$$
$$= j|B_C| \qquad 2.21$$

Example 2.15

A capacitor has a reactance of 12 Ω. Find its susceptance.

Solution

$$B_C = \frac{1}{X_C} = \frac{1}{12\ \Omega}$$
$$= 0.083\,33\ \text{S} \quad (0.083\ \text{S})$$

Example 2.16

What is the susceptance of a 10 μF capacitor at 100 Hz?

Solution

$$B_C = \frac{1}{X_L} = j\omega C$$
$$= 2\pi f C$$
$$= 2\pi\,(100\ \text{Hz})\,(10 \times 10^{-6}\ \text{F})$$
$$= 0.006283\ \text{S} \quad (6.283\ \text{mS})$$

INDUCTIVE REACTANCE

If an AC source drives an inductor, the inductor exhibits a resistance called *inductive reactance*. The current entering the inductor creates a magnetic field. The magnetic field cannot be created instantly, so there is a resistance to current flow. Energy is stored in the magnetic field. When the current decreases, the magnetic field reduces and gives up the energy stored in the magnetic field. The change in the magnetic field creates an electrical current. This current flows back into the generator. Energy is not dissipated in this process but only stored temporarily in the magnetic field.

The resistance to the magnetic field buildup is what gives the inductor its reactance (i.e., AC resistance). Inductive reactance carries units of ohms (Ω) and is

$$X_L = 2\pi f L = \omega L \qquad 2.22$$

Figure 2.16 shows an AC source charging and discharging an inductor. The current and voltage are 90° out of phase as shown in Fig. 2.17. The current lags the voltage. When the voltage is initially applied, the current is 0 A (lags) and there is no magnetic field. The current begins to flow as it builds up the magnetic field. The current relative to the excitation voltage in phasor notation carries units of amps (A) and is

$$i = i_{\text{magnitude}}\angle{-90°} \qquad 2.23$$

This is shown in Fig. 2.18.

Inductive reactance is similar to capacitive reactance; both have units of ohms and follow Ohm's law and Kirchhoff's laws. The difference between resistance and reactance is that inductive reactance and capacitive reactance store energy. Resistors do not store energy and have no time dependence. Resistors dissipate power; inductors and capacitors do not.

Complex algebra is a convenient method of analyzing circuits with inductance.

Inductive reactance magnitude is defined as

$$|X_C| = 2\pi f L \qquad 2.24$$

The inductive phase angle is

$$X_C = j\omega L = |X_C|\angle{90°}$$
$$= j|X_L| \qquad 2.25$$

Example 2.17

A source operates at 1000 Hz, driving a 22 mH inductor. Find the inductive reactance. How does the reactance change if the source frequency is changed to 10 kHz? To 1.0 MHz?

Figure 2.16 Charging and Discharging an Inductor

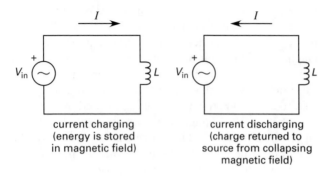

current charging
(energy is stored
in magnetic field)

current discharging
(charge returned to
source from collapsing
magnetic field)

Figure 2.17 Voltage and Current into an Inductor

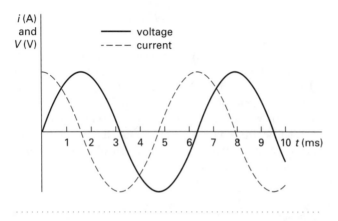

Figure 2.18 Inductive Voltage and Current Phasor Diagram

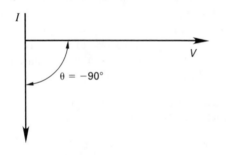

Solution

$$X_L = \omega L = 2\pi f L$$

$$= 2\pi\,(1000\text{ Hz})\,(22\text{ mH})\left(\frac{1\text{ H}}{1000\text{ mH}}\right)$$

$$= 138.23\ \Omega \quad (138\ \Omega)$$

$$X_L = 2\pi f L$$

$$= 2\pi\,(10\,000\text{ Hz})\,(22\text{ mH})\left(\frac{1\text{ H}}{1000\text{ mH}}\right)$$

$$= 1382.3\ \Omega \quad (1.38\text{ k}\Omega)$$

If the frequency is increased by a factor of 10, the reactance increases by a factor of 10.

$$X_L = 2\pi f L$$

$$= 2\pi(1.0 \times 10^6\text{ Hz})(22 \times 10^{-3}\text{ H})$$

$$= 138.23\text{ k}\Omega \quad (138\text{ k}\Omega)$$

Example 2.18

A 100 mH inductor is across the power grid with an rms voltage of 120 V at 60 Hz. What is the reactance of the inductor? What is the current?

Solution

$$|X_L| = 2\pi f L$$

$$= 2\pi\,(60\text{ Hz})\,(0.1\text{ H})$$

$$= 37.7\ \Omega$$

Given the grid rms voltage of 120 V,

$$|i| = \left|\frac{V}{X_C}\right| = \frac{120\text{ V}}{37.7\ \Omega}$$

$$= 3.183\text{ A}$$

$$i = 3.183\text{ A}\angle 90°$$

Inductive Admittance and Susceptance

Inductive susceptance is the reciprocal of inductive reactance. The variable for inductive susceptance is B and the units are siemens. Inductive reactance has a phase angle of 90°, or $j\,|X_L|$. Therefore, the inductive susceptance is

$$B_L = \frac{1}{|X_L|\ \angle 90°}$$

$$= \left|\frac{1}{X_L}\right|\ \angle - 90° \qquad 2.26$$

This is the opposite of capacitive susceptance.

Example 2.19

What is the susceptance of a 20 mH inductor at 1000 Hz?

Solution

$$B_L = \frac{1}{X_L} = \frac{1}{2\pi f L}$$

$$= \frac{1}{2\pi(1000\text{ Hz})(0.02\text{ H})}$$

$$= 0.00796\text{ S} \quad (8\text{ mS})$$

Example 2.20

A 50 mH inductor operates at 500 Hz. What is its susceptance?

Solution

$$|B_L| = \frac{1}{X_L} = \left| \frac{1}{j\omega L} \right|$$
$$= \frac{1}{2\pi f L}$$
$$= \frac{1}{2\pi (500 \text{ Hz}) (50 \times 10^{-3} \text{ H})}$$
$$= 0.0064 \text{ S} \quad (6.4 \text{ mS})$$

CAPACITIVE AND INDUCTIVE REACTANCE

Inductive reactance varies directly with frequency. Capacitive reactance varies inversely with frequency. Figure 2.19 shows how capacitive reactance and inductive reactance vary with frequency.

Figure 2.19 Inductive and Capacitive Reactance versus Frequency

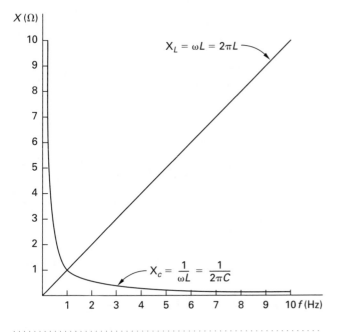

An easy pneumonic aid can be used to remember whether current leads or lags in an inductive or capacitive circuit: "ELI the ICE man." Voltage and electromotive force are the same thing, which will be represented by an "E" in the pneumonic aid. "ELI" means the voltage (E) in an inductive circuit (L) leads the current (I). "ICE" means current (I) in a capacitive circuit (C) leads the voltage (E).

SERIES RESISTOR-CAPACITOR CIRCUITS

The resistance in a series resistor-capacitor (RC) circuit is a real quantity. The series capacitive reactance is a negative, imaginary quantity. Combining the real and imaginary quantities results in impedance, represented by the variable Z.

$$
\begin{aligned}
R &= R + j0 \\
+ X_C &= 0 - \frac{j}{\omega C} \\
\hline
Z &= R - \frac{j}{\omega C}
\end{aligned}
\qquad 2.27
$$

$$\theta = \tan^{-1} \frac{X_C}{R} \qquad 2.28$$
$$|Z| = \sqrt{R^2 + |X_C|^2} \qquad 2.29$$

A series RC circuit is shown in Fig. 2.20.

Figure 2.20 Series RC Circuit

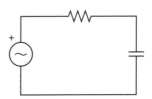

Example 2.21

A series RC circuit has a 20 Ω resistor and a capacitor of 48 Ω reactance. Find the impedance and show the phasor diagram.

Solution

$$
\begin{aligned}
R &= R + j0 &&= 20 \text{ Ω} + j0 \text{ Ω} \\
+ X_C &= 0 - \frac{j}{\omega C} &&= 0 - j48 \text{ Ω} \\
\hline
Z &= R - \frac{j}{\omega C} &&= 20 - j48 \text{ Ω}
\end{aligned}
$$

$$
\begin{aligned}
\theta &= \tan^{-1} \frac{X_C}{R} \\
&= \tan^{-1} \left(\frac{-48 \text{ Ω}}{20 \text{ Ω}} \right) \\
&= -67.38°
\end{aligned}
$$

$$
\begin{aligned}
|Z| &= \sqrt{R^2 + |X_C|^2} \\
&= \sqrt{(20 \text{ Ω})^2 + (48 \text{ Ω})^2} \\
&= 52 \text{ Ω}
\end{aligned}
$$

$$Z = 52 \text{ Ω} \angle -67.38°$$

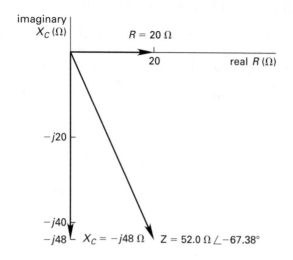

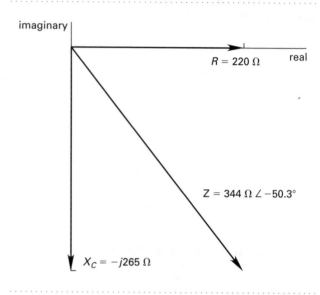

The voltage lags the current, but not by the 90° difference in Ex. 2.14, which involved a circuit without a resistor.

Example 2.22

A series RC circuit has a resistor of 220 Ω and a capacitor of 10 μF. The circuit is driven by a 12 V, 60 Hz source. What is the impedance? What is the current?

Solution

$$Z = R - j|X_C|$$

$$= R - \frac{j}{2\pi f C}$$

$$= 220 - \frac{j}{2\pi\,(60\text{ Hz})\,(10\ \mu\text{F})\left(1\times10^{-6}\ \dfrac{\text{F}}{\mu\text{F}}\right)}$$

$$= 220 - j265\ \Omega$$

$$Z = \sqrt{(220\ \Omega)^2 + (265\ \Omega)^2}\ \tan^{-1}\left(\frac{-265\ \Omega}{220\ \Omega}\right)$$

$$= 344\ \Omega\angle{-50.3°}$$

$$i = \frac{12\text{ V}}{344\ \Omega\angle{-50.3°}}$$

$$= 0.034\,84\text{ A}\angle 50.3° \quad (34.84\text{ mA}\angle 50.3°)$$

The current leads the voltage.

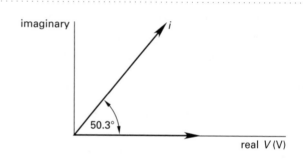

SERIES RESISTOR-INDUCTOR CIRCUITS

The resistance in a series resistor-inductor (RL) circuit is a real quantity. The inductive reactance is a positive imaginary quantity. Combining the real and imaginary quantities results in impedance, represented by the variable Z.

$$\begin{aligned} R\ &=\ R + j0 \\ +\ X_L\ &=\ 0\ +\ j\omega L \\ \hline Z\ &=\ R + j\omega L \end{aligned}$$

$$2.30$$

$$\theta = \tan^{-1}\left(\frac{2\pi L}{R}\right)$$

$$= \tan^{-1}\left(\frac{\omega L}{R}\right)$$

$$= \tan^{-1}\left(\frac{X_L}{R}\right) \qquad 2.31$$

$$|Z| = \sqrt{R^2 + |X_L|^2} \qquad 2.32$$

$$Z = |Z|\ \angle\theta \qquad 2.33$$

A series RL circuit is shown in Fig. 2.21.

Figure 2.21 Series RL Circuit

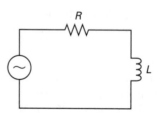

Example 2.23

A series RL circuit has a 120 Ω resistor and a 130 mH inductor. The circuit is driven by a 10 V source at 60 Hz. Find the impedance and resulting current. Draw the phasor diagram for impedance and current.

Solution

$$
\begin{aligned}
R &= R + j0 &&= 120 + j0.0\ \Omega &&= 120 + j0.0\ \Omega \\
+\ X_L &= 0 + j\omega fL &&= 0 + j2\pi(60\ \text{Hz})(0.13\ \text{H}) &&= 0 + j49.0\ \Omega \\
\hline
Z &= R + j\omega L &&= 120 + j2\pi(60\ \text{Hz})L &&= 120 + j49.0\ \Omega
\end{aligned}
$$

$$
\begin{aligned}
\theta &= \tan^{-1}\frac{\omega L}{R} \\
&= \tan^{-1}\left(\frac{2\pi(60\ \text{Hz})(0.13\ \text{H})}{120\ \Omega}\right) \\
&= \tan^{-1}\left(\frac{49\ \Omega}{120\ \Omega}\right) \\
&= 22.2°
\end{aligned}
$$

$$
\begin{aligned}
|Z| &= \sqrt{R^2 + |X|_L^2} \\
&= \sqrt{(120\ \Omega)^2 + (49\ \Omega)^2} \\
&= 130\ \Omega \\
Z &= 130\ \Omega\angle 22.2°
\end{aligned}
$$

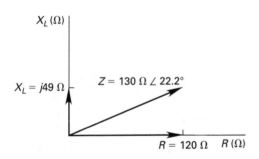

$$
\begin{aligned}
i &= \frac{10\ \text{V}}{130\ \Omega\ \angle 22.2°} \\
&= 0.077\ \text{A}\angle{-22.2°} \quad (77\angle{-22.2°}\ \text{mA})
\end{aligned}
$$

The current lags the voltage as shown.

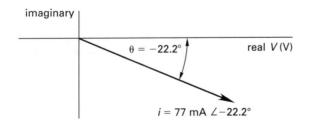

The current lags the voltage, but not by the 90° difference seen in Ex. 2.18 in the circuit without a resistor.

RESISTOR-INDUCTOR-CAPACITOR CIRCUITS

Series Resistor-Inductor-Capacitor Circuits

A series resistor-inductor-capacitor (RLC) circuit is shown in Fig. 2.22.

Figure 2.22 Series RLC Circuit

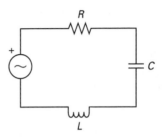

To find the total impedance, the real parts are added, and the imaginary parts are added separately.

$$
\begin{aligned}
R &= R + j0 \\
X_L &= 0 + j\omega L \\
+\ X_C &= 0 - \frac{j}{\omega C} \\
\hline
Z &= R + j\omega L - \frac{j}{\omega C} \\
&= R + j\left(\omega L - \frac{1}{\omega C}\right) \\
&= R \pm j\,|X_L - X_C| \\
&= R \pm jX &&\text{2.34}
\end{aligned}
$$

$$
\begin{aligned}
\theta &= \tan^{-1}\left(\frac{X_L - X_C}{R}\right) \\
&= \tan^{-1}\left(\frac{X}{R}\right) &&\text{2.35}
\end{aligned}
$$

$$
\begin{aligned}
|Z| &= \sqrt{R^2 + |X_L - X_C|^2} \\
&= \sqrt{R^2 + |X|^2} &&\text{2.36}
\end{aligned}
$$

Example 2.24

A series RLC circuit has a resistance of 20 Ω, an inductive reactance of 20 Ω, and a capacitive reactance of 48 Ω. Find the impedance and draw the phasor diagram showing the impedance and both reactances.

Solution

$$
\begin{aligned}
R &= R + j0 \ \Omega & &= 20 + j0 \ \Omega \\
X_L &= 0 + j\omega L & &= 0 + j20 \ \Omega \\
X_C &= 0 - \frac{j}{\omega C} & &= 0 - j48 \ \Omega
\end{aligned}
$$

$$
Z = R + j\omega L - \frac{j}{\omega C} = 20 - j28 \ \Omega
$$

$$
\theta = \tan^{-1} \frac{X_C}{R} = \tan^{-1}\left(\frac{-28 \ \Omega}{20 \ \Omega}\right)
$$

$$
= -54.46°
$$

$$
|Z| = \sqrt{R^2 + |X_T|^2}
$$

$$
= \sqrt{(20 \ \Omega)^2 + (28 \ \Omega)^2}
$$

$$
= 34.4 \ \Omega
$$

$$
Z = 34.4 \ \Omega \angle -54.46°
$$

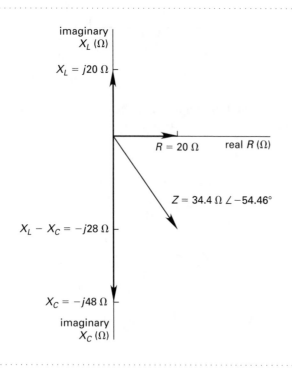

Example 2.25

The circuit in Ex. 2.24 is driven by a 100 V source. Find the voltage across all components and verify that the circuit agrees with Kirchhoff's voltage law (KVL). Find the current. Draw the vector diagram for the impedance and a second diagram showing the relationship between the input voltage and current.

Solution

$$
\begin{aligned}
Z_T &= R + j(X_L - X_C) \\
&= 20 + j(20 \ \Omega - 48 \ \Omega) \\
&= 20 - j28 \ \Omega \\
&= 34 \ \Omega \angle -54.46°
\end{aligned}
$$

Using the voltage divider rule,

$$
\begin{aligned}
v_C &= V_{\text{in}} \frac{X_C}{Z_T} \\
&= (100 \ \text{V})\left(\frac{48 \ \Omega \angle -90°}{34 \ \Omega \angle -54.46°}\right) \\
&= 141 \ \text{V} \angle(-90° - (-54.46°)) \\
&= 141 \ \text{V} \angle -35.54° \\
&= 114 - j82 \ \text{V}
\end{aligned}
$$

$$
\begin{aligned}
v_L &= V_{\text{in}} \frac{X_L}{Z_T} \\
&= (100 \ \text{V})\left(\frac{20 \ \Omega \angle 90°}{34 \ \Omega \angle -54.46°}\right) \\
&= 58.1 \ \text{V} \angle(90° - (-54.46°)) \\
&= 58 \ \text{V} \angle 144.46° \\
&= -48 + j34 \ \text{V}
\end{aligned}
$$

$$
\begin{aligned}
v_R &= V_{\text{in}} \frac{R}{Z_T} \\
&= (100 \ \text{V})\left(\frac{20 \ \Omega \angle 0°}{34 \ \Omega \angle -54.46°}\right) \\
&= 59 \ \text{V} \angle 54.46° \\
&= 34 + j48 \ \text{V}
\end{aligned}
$$

Checking results using KVL,

$$
\begin{aligned}
v_C &= 114 - j82 \ \text{V} \\
v_L &= -48 + j34 \ \text{V} \\
+ \ v_R &= 34 + j48 \ \text{V} \\
\hline
V_{\text{in}} &= 100 \pm j0.0 \ \text{V}
\end{aligned}
$$

The circuit satisfies KVL.

v_C is greater than V_{in}. This is not a mistake. KVL confirms that this seemingly impossible voltage is correct.

$$
\begin{aligned}
i = \frac{V}{Z} &= \frac{100 \ \text{V} \angle 0°}{34 \ \Omega \angle -54.46°} \\
&= 2.9 \ \text{A} \angle 54.46°
\end{aligned}
$$

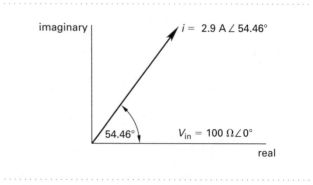

Parallel Resistor-Inductor-Capacitor Circuits

Parallel RLC circuits are best analyzed with conductance and susceptance. The combination of conductance and susceptance is admittance. Admittance is the reciprocal of impedance. Admittance units are siemens and the admittance symbol is Y.

$$Y = \frac{1}{Z} \qquad 2.37$$

$$Y = G + j\left(B_C - B_L\right) \qquad 2.38$$

Example 2.26

A parallel RLC parallel circuit has conductance of 0.05 S, inductive susceptance of 0.05 S, and capacitive susceptance of 0.09 S. What is the parallel impedance?

Solution

$$\begin{aligned}
Y &= G + j\left(B_C - B_L\right)_L \\
&= 0.05\ \text{S} + j\left(0.09\ \text{S} - 0.05\ \text{S}\right) \\
&= 0.05 + j\,0.04\ \text{S} \\
&= 0.064\ \text{S}\angle 38.66^\circ
\end{aligned}$$

$$\begin{aligned}
Z &= \frac{1}{Y} \\
&= \frac{1}{0.064\ \text{S}\angle 38.66^\circ} \\
&= 15.6\ \Omega\angle{-38.66^\circ} \\
&= 12 - j9.8\ \Omega
\end{aligned}$$

Example 2.27

A parallel circuit has a resistor of 10 Ω, a capacitor of 10 μF, and an inductor of 12 mH. The circuit is operating at 400 Hz. What is the parallel impedance?

Solution

$$\begin{aligned}
B_C &= j\omega C \\
&= j2\pi f C \\
&= j2\pi \left(400\ \text{Hz}\right)\left(10 \times 10^{-6}\ \text{F}\right) \\
&= j0.025\ \text{S}
\end{aligned}$$

$$\begin{aligned}
B_L &= \frac{1}{j\omega L} = \frac{1}{j2\pi f L} \\
&= \frac{1}{j2\pi\left(400\ \text{Hz}\right)\left(12 \times 10^{-3}\ \text{H}\right)} \\
&= -j0.033\ \text{S}
\end{aligned}$$

$$\begin{aligned}
Y &= G + j\left(B_C - B_L\right) \\
&= 0.10\ \text{S} + j\left(0.025\ \text{S} - 0.033\ \text{S}\right) \\
&= 0.10 - j\,0.008\ \text{S} \\
&= 0.10\ \text{S}\angle{-4.57^\circ}
\end{aligned}$$

$$\begin{aligned}
Z &= \frac{1}{Y} = \frac{1}{0.10\ \text{S}\angle{-4.57^\circ}} \\
&= 10\ \Omega\angle 4.57^\circ \\
&= 10 + j0.80
\end{aligned}$$

POWER FACTOR

Maximum power is delivered when voltage and current are in phase ($\theta = 0^\circ$). When they are out of phase ($\theta \neq 0^\circ$), less power will be delivered to the load. Figure 2.23 shows the power triangle, a graphic representation of the relationship between real power, P, reactive power, Q, apparent power, S, and the power angle, θ.

Figure 2.23 Power Triangle

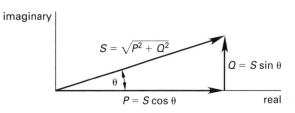

The *real power* the power actually delivered to the resistive load, is measured in watts. The total power sent down the transmission line is the *apparent power*, measured in volt-amperes (V·A). The apparent power is the hypotenuse of the power triangle.

The *reactive power* just uses the transmission facilities and does not contribute anything to the power to the load. Reactive power has units of volt-amperes-reactive (VAR), not watts, indicating no power is delivered.

The *power factor* is the cosine of the power angle (cos θ). The closer the power factor is to 1.0, the more power given to the load and the reactive power is minimized. Power companies are quite concerned about keeping the power factor close to 1.0. The transmission system is highly inductive, with many transformers, motors, and long wires that tend to reduce the power factor. Capacitors are switched onto the power grid to bring the power factor closer to 1.0.

Example 2.28

A generator outputs an rms voltage of 220 V to a load. The rms current is 1.4 A, leading the voltage by 12°. How much power is delivered to the load? What is the reactive power? What is the apparent power?

Solution

$$\begin{aligned}
|P| &= VI\cos\theta \\
&= \left(220\ \text{V}\right)\left(1.4\ \text{A}\right)\cos 12^\circ \\
&= 301\ \text{W} \\
|Q| &= VI\sin\theta \\
&= \left(220\ \text{V}\right)\left(1.4\ \text{A}\right)\sin 12^\circ \\
&= 64\ \text{VAR}
\end{aligned}$$

$$|S| = \sqrt{P^2 + Q^2}$$
$$= \sqrt{(301 \text{ W})^2 + (64 \text{ VAR})^2}$$
$$= 308 \text{ V}\cdot\text{A}$$

RESONANCE

Some mechanical systems that have resonances are the spring, mass-and-dashpot system, and bridge and automobile suspension systems.

Some electrical systems also have resonances. In electrical systems, resonance occurs when inductive reactance equals capacitive reactance. These electrical resonances are used to select radio stations and control oscillator frequencies. Mechanical and electrical resonance is described by a second order differential equation, and so it is called a second order system. No calculus is needed to understand resonance basics.

An electrical resonance is made by connecting an inductor, capacitor, and resistor in series as shown in Fig. 2.24.

Figure 2.24 Series RLC Circuit

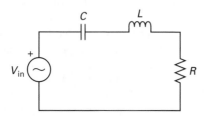

In such a resonant circuit, the equation for impedance becomes

$$|Z| = \sqrt{R \pm (jX_L - jX_C)}$$
$$= \sqrt{R \pm j(X_L - X_C)} \qquad 2.39$$

In the complex plane, the impedance equation can be shown as in Fig. 2.25.

In a series RLC circuit, such as the one in Fig. 2.24, maximum power is sent to the load by keeping the power factor as close to 1.0 as possible. If the circuit is purely resistive, that is $|X_L - X_C| = 0 \ \Omega$, then the impedance equals the resistance, and impedance is at its minimum value. If the impedance is at minimum, then voltage and current are at maximum and the power factor is 1.0 (i.e., $\cos\theta = 1.0$). In this situation, maximum power is transferred to the resistor (the load).

$$I = \frac{V}{Z} = \frac{V}{R + j(X_L - X_C)}$$

Figure 2.25 Series RLC Impedance Phasor Diagram

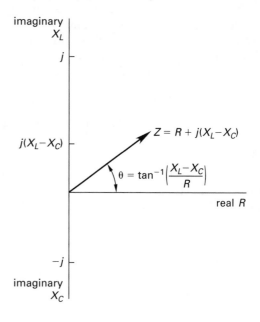

$$= \frac{V}{R + j0}$$
$$= \frac{V}{R} \qquad 2.40$$
$$V_R = IR \qquad 2.41$$

If the denominator's reactive portion is anything other than zero magnitude, the current is reduced, and the power delivered is reduced along with the current.

Resonance occurs when inductive reactance equals capacitive reactance, which can be written as $|X_L - X_C| = 0 \ \Omega$ or $|X_L| = |X_C|$. The formula for resonance is derived as follows.

If resonance occurs,

$$X_L = 2\pi f L = \omega f L \qquad 2.42$$
$$X_C = \frac{1}{2\pi f C} = \frac{1}{\omega C} \qquad 2.43$$

Since $X_L = X_C$ at resonance,

$$\omega L = \frac{1}{\omega C} \qquad 2.44$$
$$\omega_0^2 = \frac{1}{LC} \qquad 2.45$$
$$\omega_0 = \frac{1}{\sqrt{LC}} \qquad 2.46$$
$$f_0 = \frac{1}{2\pi\sqrt{LC}} \qquad 2.47$$

Example 2.29

A series RLC circuit has an inductance of 10 mH and a capacitance of 1 μF. Find the resonant frequency.

Solution

$$\omega = \frac{1}{\sqrt{LC}}$$

$$= \frac{1}{\sqrt{(0.01 \text{ H})(1 \times 10^{-6} \ \mu\text{F})}}$$

$$= 10\,000 \text{ rad/s}$$

$$f = \frac{\omega}{2\pi} = \frac{10\,000 \ \frac{\text{rad}}{\text{s}}}{2\pi}$$

$$= 1592 \text{ Hz}$$

QUALITY FACTOR

Tuned resonant RLC circuits are used in radio and television receivers to select only one station at a time. Figure 2.26 shows how the impedance and power transfer change as the source frequency varies.

Figure 2.26 RLC Circuit Resonance

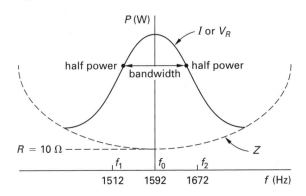

Station signals that use frequencies outside the bandwidth of the selected resonant frequency, f_0, will not be as strong as the station within the bandwidth. Bandwidth, represented by BW, is the difference between the half-power points, f_2 and f_1. The *quality factor*, Q, is a measure of the circuit bandwidth or selectivity.

$$\text{BW} = f_2 - f_1 = \frac{f_0}{Q} \qquad 2.48$$

$$Q = \frac{f_0}{\text{BW}} \qquad 2.49$$

$$f_1 = f_0 - \frac{\text{BW}}{2} \qquad 2.50$$

$$f_2 = f_0 + \frac{\text{BW}}{2} \qquad 2.51$$

These equations are valid only if the quality factor is greater than or equal to 10. Q is independent of frequency.

Example 2.30

The tuned resonant RLC circuit in Fig. 2.26 has a resistance of 10 Ω, an inductance of 10 mH, and capacitance of 1 μF. From Ex. 2.29, the resonant frequency is 1592 Hz. Find the quality factor, bandwidth, and half-power points of the circuit.

Solution

$$Q = \frac{f_0}{\text{BW}} = \frac{f_0}{f_2 - f_1}$$

$$= \frac{\omega_0 L}{R}$$

$$= \frac{2\pi f_0 L}{R}$$

$$= \frac{2\pi (1592 \text{ Hz}) (0.01 \text{ H})}{10 \ \Omega}$$

$$= 10$$

$$Q = \frac{f_0}{\text{BW}} = \frac{f_0}{f_2 - f_1}$$

$$= \frac{1}{\omega_0 RC}$$

$$= \frac{1}{2\pi f_0 RC}$$

$$= \frac{1}{2\pi (1592 \text{ Hz}) (10 \ \Omega)(1 \ \mu\text{F})}$$

$$= 10$$

$$\text{BW} = f_2 - f_1 = \frac{f_0}{Q}$$

$$= \frac{1592 \text{ Hz}}{10}$$

$$= 159.2 \text{ Hz}$$

$$f_1 = f_0 - \frac{\text{BW}}{2}$$

$$= 1592 \text{ Hz} - \frac{159.2 \text{ Hz}}{2}$$

$$= 1512 \text{ Hz}$$

$$f_2 = f_0 + \frac{\text{BW}}{2}$$

$$= 1592 \text{ Hz} + \frac{159.2 \text{ Hz}}{2}$$

$$= 1672 \text{ Hz}$$

MAXIMUM POWER TRANSFER

With DC circuits, the maximum power transferred to a load occurs when source resistance and load resistance are equal. In AC circuits, the maximum power transfer occurs when the source impedance equal the load resistance conjugate. That is, maximum power transfer occurs when

$$Z_{\text{source}} = Z_{\text{load}}^*$$

The inductive and capacitive reactances cancel.

Example 2.31

A load has an impedance of

$$Z = 50 + j10 \ \Omega$$

What source impedance is required for maximum power transfer?

Solution

$$Z_{\text{source}} = Z_{\text{load}}^*$$

The load impedance is

$$Z = 50 + j10 \ \Omega$$

The conjugate is

$$50 - j10 \ \Omega$$

This is the required source impedance. Radio and television transmitters have inductance-capacitance (LC) tuners to match the transmitter output to the transmission line and antenna.

TRANSFORMERS

A transformer has two inductors placed close enough together that their two magnetic fields interact. A power transformer with an iron core is shown in Fig. 2.27. Iron is a ferromagnetic material which readily conducts magnetic fields. In equations involving transformers, N_P represents the number of wire turns on the primary coil of the transformer. The primary coil (or winding) is connected to the source. N_S represents the number of turns on the secondary coil. The secondary coil is connected to the load. V_P is the primary coil voltage, and I_P is the primary coil current. V_S is the secondary coil voltage, and I_S is the secondary coil current.

Figure 2.27 Power Transformer

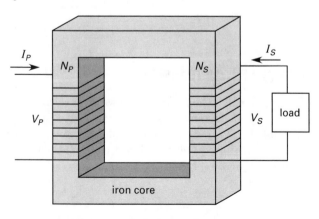

Transformers are used to increase (step up) voltage or decrease (step down) voltage. If the output voltage is increased, the output current must decrease proportionally. If the output voltage is reduced, the output current must be increased proportionally. In other words, the power output must equal the power input, ignoring losses. A transformer does not create power, and an ideal transformer does not lose power.

Transformers are also used to isolate portions of a circuit. There is no electrical connection between the primary and secondary coils. The only interaction is through magnetic fields.

An input current creates a changing magnetic field in a transformer. In the transformer shown in the illustration, the iron ferromagnetic core readily conducts the primary coil's magnetic field to the secondary coil. The secondary voltage is induced by the changing magnetic field. DC does not have a changing magnetic field. A transformer must be driven by AC.

Transformer output voltage is proportional to the ratio of secondary coil turns to primary coil turns. The following equations are based on the principles that power output must equal power input and that power equals the voltage multiplied by the current.

$$\frac{V_S}{V_P} = \frac{N_S}{N_P} \qquad 2.52$$

$$V_S = \frac{N_S}{N_P} V_P \qquad 2.53$$

$$\frac{I_S}{I_P} = \frac{N_P}{N_S} \qquad 2.54$$

$$I_S = \frac{N_P}{N_S} I_P \qquad 2.55$$

Example 2.32

An ideal transformer has 280 secondary coil turns, and 70 primary turns. If the input voltage is 230 V and the input current is 1.0 A, what is the output voltage, output current, and power? Compare the output power to the input power.

Solution

$$V_S = \frac{N_S}{N_P} V_P$$
$$= \left(\frac{280 \text{ turns}}{70 \text{ turns}} \right) (230 \text{ V})$$
$$= 920 \text{ V}$$
$$I_S = \frac{N_P}{N_S} I_P$$
$$= \left(\frac{70 \text{ turns}}{280 \text{ turns}} \right) (1.0 \text{ A})$$
$$= 0.25 \text{ A}$$

$$P_P = V_P I_P$$
$$= (230 \text{ V}) (1.0 \text{ A})$$
$$= 230 \text{ W}$$
$$P_S = V_S I_S$$
$$= (920 \text{ V}) (0.25 \text{ A})$$
$$= 230 \text{ W}$$

The input power and output power are identical for this ideal transformer.

Example 2.33

A transformer has 200 primary turns, 400 secondary turns, and an input voltage of 115 V. What is the output voltage? If the load is 100 Ω, what is the output and input current?

Solution

$$V_S = \frac{N_S}{N_P} V_P$$
$$= \left(\frac{400 \text{ turns}}{200 \text{ turns}} \right) (115 \text{ V})$$
$$= 230 \text{ V}$$

The secondary current is

$$I_S = \frac{V_S}{R} = \frac{230 \text{ V}}{100 \text{ }\Omega}$$
$$= 2.3 \text{ A}$$

The primary current is

$$I_S = \frac{N_P}{N_S} I_P$$
$$I_P = \frac{N_S}{N_P} I_S$$
$$= \left(\frac{400 \text{ turns}}{200 \text{ turns}} \right) (2.3 \text{ A})$$
$$= 4.6 \text{ A}$$

Maximum Power Transfer in a Transformer

Another use of transformers is to maximize power transfer. If a source's Thevenin resistance and the load resistance cannot be made equal, maximum power transfer will not happen. The turns ratios of transformers can be used to match source resistance to load resistance.

Figure 2.28 Maximum Power Transfer Using a Transformer

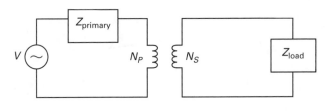

The formula for maximum power transfer is

$$R_{\text{source}} = \left(\frac{N_P}{N_S} \right)^2 R_{\text{load}} \qquad 2.56$$

Example 2.34

A transformer with 200 primary turns and 100 secondary turns drives a 50 Ω load. What source resistance will match this load?

Solution

$$R_{\text{source}} = \left(\frac{N_P}{N_S} \right)^2 R_{\text{load}}$$
$$= \left(\frac{200 \text{ turns}}{100 \text{ turns}} \right)^2 (50 \text{ }\Omega)$$
$$= (2)^2 (50 \text{ }\Omega)$$
$$= 200 \text{ }\Omega$$

Example 2.35

An engineer wants to match a loudspeaker with a resistance of 16 Ω to a 256 Ω source. What is the necessary transformer turns ratio?

Solution

$$R_{\text{source}} = \left(\frac{N_P}{N_S} \right)^2 R_{\text{load}}$$
$$\frac{N_P}{N_S} = \sqrt{\frac{R_{\text{source}}}{R_{\text{load}}}}$$
$$= \sqrt{\frac{256 \text{ }\Omega}{16 \text{ }\Omega}}$$
$$= \sqrt{16}$$
$$= 4$$

Transformer Efficiency and Energy Loss

Real transformers have energy losses. The winding wires lose energy through ohmic resistance (i.e., losses described by $P = I^2 R$). The magnetic field sets up eddy currents in the iron core which absorb energy. In addition, the constantly varying input current in a transformer continuously changes the magnetic domains' orientations and uses energy. This process is called *hysteresis*, and it is inherent in all magnetic materials.

Efficiency is power delivered divided by power input. It is usually expressed as a percentage and given the symbol η.

$$\eta = \left(\frac{P_{\text{out}}}{P_{\text{in}}} \right) \times 100\% \qquad 2.57$$

As the input frequency increases, these losses increase to the point that iron core transformers become extremely inefficient. At these higher frequencies, the core material of choice is ferrite, an epoxy impregnated with an iron alloy powder. The small iron particles in ferrite have smaller eddy-current losses than a solid iron core. These types of transformers are rarely used as power transformers.

At frequencies above 2 MHz, the ferrous core is eliminated entirely from the transformer and an air core is used in its place. An air core transformer does not have any ferrous magnetic materials.

Example 2.36

A transformer operates with inputs of 117 V and 2.2 A, and it has a 12.6 V output at 19.0 A. What is its efficiency?

Solution

$$\eta = \left(\frac{P_{\text{out}}}{P_{\text{in}}}\right) \times 100\%$$

$$= \frac{(12.6\ \text{V})\,(19.0\ \text{A})}{(117\ \text{V})\,(2.2\ \text{A})} \times 100\%$$

$$= 93.0\%$$

Transformer Dot Notation

Transformer output voltages can be either in phase or 180° out of phase with the input voltage, depending on the direction of secondary coil windings. A dot on both the transformer and the schematic drawing indicates the relative phases, as shown in Fig. 2.29.

Figure 2.29 Transformer Dot Notation

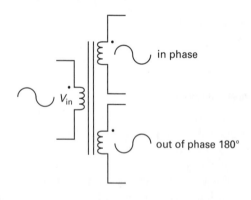

AC GENERATOR

An AC generator produces voltage through the interaction of a magnet moving past stationary coils. A simplified AC generator is shown in Fig. 2.30. By Faraday's law, the changing magnetic field induces a voltage. The voltage, e, equals the number of turns on the coil times the rate of change of the magnetic field.

$$e = -N\,\frac{\Delta\phi}{\Delta t} \qquad\qquad 2.58$$

The north pole of the magnet induces a positive voltage, and the south pole induces a negative voltage. Generators are described further in the next chapter.

Figure 2.30 AC Generator

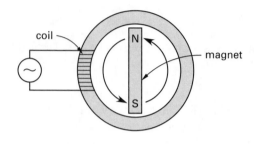

PRACTICE PROBLEMS

1. An AC signal

    (A) is one that varies with time
    (B) changes polarity with time
    (C) changes between a negative value and zero
    (D) changes between a positive value and zero

2. Which of the following waveforms is an AC signal?

    (A)

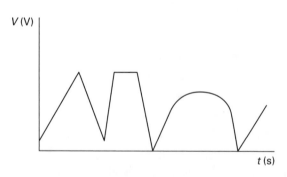

    (B)

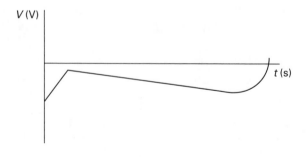

(C)

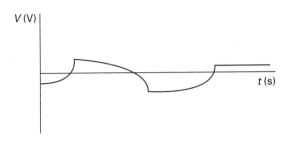

(D)

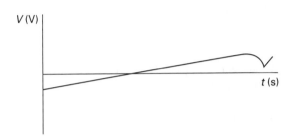

3. Find the period and frequency for the waveform.

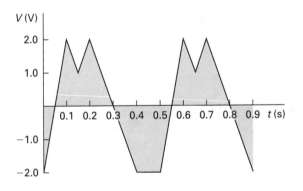

(A) $T = 0.25$ s, $f = 4.0$ Hz
(B) $T = 0.50$ s, $f = 2.0$ Hz
(C) $T = 0.55$ s, $f = 1.8$ Hz
(D) $T = 0.90$ s, $f = 9.0$ Hz

4. Find the peak-to-peak voltage for the waveform shown in Prob. 3.

(A) -2.0 V
(B) 1.0 V
(C) 2.0 V
(D) 4.0 V

5. What is most nearly the rms current for the waveform?

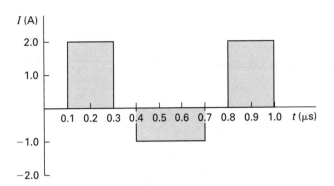

(A) 1.20 A
(B) 1.25 A
(C) 1.50 A
(D) 2.80 A

6. What is most nearly the average current for the waveform shown in Prob. 5?

(A) 0.143 A
(B) 0.201 A
(C) 1.29 A
(D) 1.49 A

7. What is most nearly the rms voltage for an AC sine wave with a peak voltage of 12 V? The waveform is symmetrical about the 0 V axis.

(A) 0.0 V
(B) 8.5 V
(C) 17 V
(D) 34 V

8. An AC sine wave is symmetrical about the 0 V axis. The negative peak value is 188 V. What is most nearly the average voltage?

(A) -100 V
(B) 0 V
(C) 100 V
(D) 300 V

9. An AC sine wave has an rms voltage of 12.6 V. The waveform is symmetrical about the 0 V axis. What is most nearly the peak negative voltage?

(A) -18 V
(B) 8.9 V
(C) 18 V
(D) 37 V

10. Find the phase shift between the two waveforms in the illustration. Waveform A is the reference. (Several zero points are given to aid the solution.)

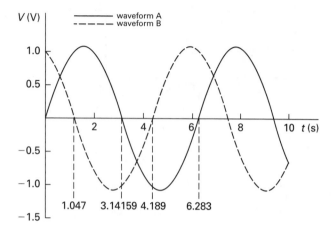

(A) Waveform B lags waveform A by 45°.
(B) Waveform B lags waveform A by 60°.
(C) Waveform B leads waveform A by 90°.
(D) Waveform B leads waveform A by 120°.

11. The peak voltage of the waveforms shown in Prob. 10 is 1.1 V. Write the time equation for waveform B.

(A) $v(t) = 1.1 \text{ V} \sin(t + 60°)$
(B) $v(t) = 1.1 \text{ V} \sin(t - 60°)$
(C) $v(t) = 1.1 \text{ V} \sin t$
(D) $v(t) = 1.0 \text{ V} \sin(t + 60°)$

12. Convert 120 Hz to the approximate radian frequency.

(A) 120 rad/s
(B) 380 rad/s
(C) 750 rad/s
(D) 750 Hz

13. A 120 V sinusoidal waveform has a frequency of 50 Hz. What is most nearly the phase angle at 8 ms? Assume the phase angle is 0° at a time of 0 ms (i.e., no phase shift).

(A) 90°
(B) 140°
(C) 170°
(D) 290°

14. What is most nearly the reactance of a 2 μF capacitor at 200 Hz?

(A) 0.40 mΩ
(B) 2.4 mΩ
(C) 400 Ω
(D) 2500 Ω

15. Given $v(t) = (222 \text{ V}) \sin 22\,000t$ driving a 3 nF capacitor, what is most nearly the AC current?

(A) 14.7 mA$\angle-90°$
(B) 14.7 mA$\angle90°$
(C) 92.1 mA$\angle90°$
(D) 15 200 mA$\angle90°$

16. What is most nearly the reactance of a 2 mH inductor at 2000 Hz?

(A) 0.040 Ω
(B) 4.0 Ω
(C) 8.0 Ω
(D) 25 Ω

17. Given $v(t) = (29 \text{ V}) \sin 2400t$ driving a 3 mH inductor, what is most nearly the AC current?

(A) 0.6 A$\angle-90°$
(B) 0.6 A$\angle90°$
(C) 4 A$\angle-90°$
(D) 4 A$\angle90°$

18. A series RLC circuit has values of $R = 10 \ \Omega$, $X_L = j6.0 \ \Omega$, and $X_C = -j9 \ \Omega$. The circuit is driven by a sinusoidal source of $v(t) = 12$ V. What is most nearly the peak current? Include the correct phase angle.

(A) 0.67 A$\angle-56°$
(B) 1.1 A$\angle0°$
(C) 1.1 A$\angle17°$
(D) 10 A$\angle0°$

19. A sine wave generator with an output of (165 V) $\sin 3000t$ is driving a series RLC circuit with the values $R = 51 \ \Omega$, $C = 10 \ \mu$F, and $L = 20$ mH. What is most nearly the peak current?

(A) 2.0 A$\angle0°$
(B) 2.0 A$\angle-28°$
(C) 2.8 A$\angle-28°$
(D) 4.0 A$\angle-28°$

20. An LC circuit has values of $L = 22$ μH and $C = 100$ pF. What is most nearly the resonant frequency in hertz?

(A) 2.9×10^{-7} Hz
(B) 0.021 Hz
(C) 3.4×10^6 Hz
(D) 21×10^6 Hz

21. A power system operates at an rms voltage of 440 V and an rms current of 3.5 A. The current lags the voltage by 12°. Most nearly how much power is delivered to the load?

(A) 125 W
(B) 320 W
(C) 1510 W
(D) 1540 W

22. A power system operates at an rms voltage of 440 V and an rms current of 3.5 A. The current lags the voltage by 12°. What is most nearly the reactive power?

(A) 320 VAR
(B) 320 W
(C) 1510 W
(D) 1540 VAR

23. A series RLC resonant circuit has resonance at 100 MHz, a lower bandwidth limit of 96.0 MHz, and an upper bandwidth limit of 104.1 MHz. What is most nearly the circuit's quality factor, Q?

(A) 8.1
(B) 12
(C) 24
(D) 25

24. An AC source has an output Thevenin impedance of $75 + j34$ Ω. What should the load impedance be for maximum power transfer?

(A) $75 - j34$ Ω
(B) 75 Ω
(C) $75 + j34$ Ω
(D) 82 Ω

25. A transformer has 220 primary turns and 660 secondary turns. The AC input voltage is 117 V. What is most nearly the output voltage?

(A) 0.0 V
(B) 39 V
(C) 170 V
(D) 350 V

26. A transformer has 220 primary turns and the secondary has 23 turns. What is most nearly the output voltage if the DC input is 234 V?

(A) 0.0 V
(B) 25 V
(C) 230 V
(D) 2200 V

27. A transformer is used to match a source resistance of 330 Ω to an 8 Ω speaker. What is most nearly the transformer turns ratio?

(A) 6.4 primary to secondary
(B) 6.4 secondary to primary
(C) 41 primary to secondary
(D) 41 secondary to primary

28. A 100% efficient transformer has an AC input of 117 V, an AC output of 234 V, and an input current of 2.2 A. What is most nearly the output current?

(A) 1.1 A
(B) 2.0 A
(C) 2.2 A
(D) 4.4 A

29. A transformer has an rms voltage input of 440 V and an rms current input of 4.0 A. The secondary load dissipates 1600 W. What is most nearly the percentage efficiency?

(A) 0.90
(B) 90%
(C) 100%
(D) 110%

30. A parallel circuit has a resistor of 20 Ω, a capacitor of 20 μF, and an inductor of 20 mH. The circuit is operating at 60 Hz. What is most nearly the parallel impedance?

(A) 0.19 Ω
(B) $2.2 - j6.3$ Ω
(C) $2.7 + j6.8$ Ω
(D) 5.2 Ω

SOLUTIONS

1. The definition of AC is alternating polarities (continually changing from positive to negative and back to positive).

Answer is B.

2. The waveform in the illustration for option (C) repeatedly crosses the 0 V amplitude line, a pattern that shows a polarity change and marks the wave as an AC signal.

Note: The illustrations in other options do not show AC signals. Options (A) and (B) change amplitude but do not cross the 0 V amplitude line. Option (D) crosses the 0 V line only once.

Answer is B.

3. Period is defined as the time between the point when a single cycle of a waveform begins and the point when it ends. The waveform in the illustration begins to repeat itself after 0.5 s. Thus the period is 0.5 s.

The frequency is

$$f = \frac{1}{T} = \frac{1}{0.5 \text{ s}}$$
$$= 2.0 \text{ Hz}$$

Any two points can be used for reference for the period. These do not necessarily need to be crossing points on the 0 s line. For instance, other acceptable beginning and end points are 0.8 s and 0.3 s, 0.6 s and 0.1 s, or 0.7 s and 0.2 s.

Note: The waveforms in options (A), (C), and (D) do not repeat themselves at the proper times.

Answer is B.

4. The peak-to-peak voltage is
$$V_{PP} = 2.0 \text{ V} - (-2.0 \text{ V})$$
$$= 4.0 \text{ V}$$

Note: Option (A) is the negative peak voltage. Option (B) is the dip at 0.15 s and 0.65 s. Option (C) is wrong because it represents the positive peak voltage.

Answer is D.

5. The rms current is
$$I_{rms} = I_{mean}$$
$$= \sqrt{\frac{(2 \text{ A})^2 (0.2 \text{ }\mu\text{s}) + (-1 \text{ A})^2 (0.3 \text{ }\mu\text{s})}{0.7 \text{ }\mu\text{s}}}$$
$$= 1.254 \text{ A} \quad (1.25 \text{ A})$$

Note: Option (A) is incorrect because it merely counts the squares above and below 0 A. In option (C), the period is incorrect. Option (D) leaves information out of the equation.

Answer is B.

6. The average current is

$$I_{avg} = \frac{A_{total}}{T}$$
$$= \frac{(2 \text{ A}) (0.2 \text{ }\mu\text{s}) + (-1 \text{ A}) (0.3 \text{ }\mu\text{s})}{0.7 \text{ }\mu\text{s}}$$
$$= 0.143 \text{ A}$$

Note: Option (B) divides the peak current by the square root of 2. Option (C) only counts the area of the squares (10 A·μs) and divides by the period. Option (D) represents the rms current.

Answer is A.

7. The rms voltage for the sine wave is

$$\frac{V_P}{\sqrt{2}} = \frac{12 \text{ V}}{\sqrt{2}}$$
$$= 8.485 \text{ V} \quad (8.5 \text{ V})$$

Note: Option (A) is the average voltage. Option (C) multiplies the peak voltage by the square root of 2 instead of dividing, and option (D) does the same thing but also uses an incorrect peak voltage of 24 V.

Answer is B.

8. The average value of any sine wave that is symmetrical around the 0 V axis is 0 V. Any other value is incorrect.

Answer is B.

9. The negative peak voltage is

$$V_{P-} = -\sqrt{2}V_{rms} = -\sqrt{2} (12.6 \text{ V})$$
$$= -17.82 \text{ V} \quad (-18 \text{ V})$$

Note: Option (B) divides the rms voltage by the square root of 2 instead of multiplying, and mixes up peak and rms voltages. Option (C) is the positive peak voltage. Option (D) is the peak-to-peak voltage.

Answer is A.

10. Waveform B leads waveform A. The period is 2π s. The phase shift is

$$\theta = \frac{\Delta t}{T} (360°)$$
$$= \left(\frac{3.14159 \text{ s} - 1.047 \text{ s}}{6.283 \text{ s}} \right) (360°)$$
$$= 120°$$

Note: Options (A) and (D) use the wrong axis crossing point. In option (B), waveform B lags instead of leads.

Answer is D.

11. The phase shift is found by

$$\theta = \frac{\Delta t}{T}(360°) = \left(\frac{1.047\ \text{s}}{6.283\ \text{s}}\right)(360°) = 60°$$

The phase is positive because it crosses to the right of the zero time axis.

Thus, the time equation for waveform B is

$$v(t) = 1.1\ \text{V}\sin(t + 60°)$$

Note: Option (B) is an incorrect phase, option (C) neglects the phase, and option (D) uses an incorrect amplitude.

Answer is A.

12. Convert 120 Hz to radian frequency.

$$\omega = 2\pi f = 2\pi(120\ \text{Hz})$$
$$= 754\ \text{rad/s}\quad(750\ \text{rad/s})$$

Note: Option (A) does not convert the frequency, option (B) uses the incorrect equation of $2\pi f = \pi(120\ \text{Hz}) = 377\ \text{rad/s}$, and option (D) uses the wrong units of hertz in the answer.

Answer is C.

13. The phase angle can be calculated as follows.

$$T = \frac{1}{f} = \frac{1}{50\ \text{Hz}}$$
$$= 20\ \text{ms}$$
$$\theta = \frac{\Delta t}{T}(360°) = \left(\frac{8\ \text{ms}}{20\ \text{ms}}\right)(360°)$$
$$= 144°\quad(140°)$$

Note: Option (A) picks the wrong angle. Option (C) uses 60 Hz instead of 50 Hz. Option (D) uses 180° instead of 360°.

Answer is B.

14. The reactance of the capacitor is

$$X_C = \frac{1}{2\pi f C} = \frac{1}{2\pi(2\ \mu\text{F})(200\ \text{Hz})}$$
$$= 398\ \Omega\quad(400\ \Omega)$$

Note: Option (A) did not use the radian frequency, nor did it invert the term. Option (B) did not take the inverse. Option (D) did not use the radian frequency.

Answer is C.

15. The AC current can be calculated as follows.

$$X_C = \left(\frac{1}{\omega C}\right)^{-1}$$
$$= \frac{1}{\left(22\,000\ \dfrac{\text{rad}}{\text{s}}\right)(3\times 10^{-9}\ \text{F})}$$
$$= 15\,151.5\ \Omega\angle{-90°}$$
$$i_C = \frac{v}{X_C} = \frac{222\ \text{V}}{15\,151.5\ \Omega\angle{-90°}}$$
$$= 0.014652\ \text{A}\quad(14.652\ \text{mA}\angle 90°)$$

Note: Option (A) uses the incorrect phase angle. Option (C) uses hertz instead of radian frequency. Option (D) is the capacitive reactance only.

Answer is B.

16. The reactance of the inductor is

$$X_L = 2\pi f L$$
$$= 2\pi(2000\ \text{Hz})(2\times 10^{-3}\ \text{H})$$
$$= 25.1\ \Omega\quad(25\ \Omega)$$

Note: Option (A) is the inverse of the correct answer, inverted using the capacitive reactance equation. In option (B), hertz is used instead of the radian frequency. In option (C), π is left out of the equation.

Answer is D.

17. The AC current can be calculated as follows.

$$X_L = \omega L$$
$$= \left(2400\ \dfrac{\text{rad}}{\text{s}}\right)(3.0\times 10^{-3}\ \text{H})$$
$$= 7.2\ \Omega$$
$$Z_L = 7.2\ \Omega\angle 90°$$
$$i_L = \frac{v}{Z_L} = \frac{29\ \text{V}}{7.2\ \Omega\angle 90°}$$
$$= 4.028\ \text{A}\angle{-90°}\quad(4.0\ \text{A}\angle{-90°})$$

Note: In option (A), hertz was used instead of radian frequency. Option (B) uses the wrong phase and uses hertz instead of radian frequency. Option (D) uses an incorrect phase.

Answer is C.

18. The peak current can be calculated as follows.

$$Z = R + j\,(X_L - X_L)$$
$$= 10\ \Omega + j\,(6.0\ \Omega - 9.0\ \Omega)$$
$$= 10 - j3.0\ \Omega$$
$$= 10.440\ \Omega\angle{-17.0°}$$
$$i_P = \frac{v}{Z} = \frac{12\ \text{V}\angle 0°}{10.440\ \Omega\angle{-17.0°}}$$
$$= 1.1149\ \text{A}\angle 17.0° \quad (1.1\ \text{A}\angle 17°)$$

Note: In option (A), the two reactances are added instead of subtracted. Option (B) does not include the phase angle. Option (D) is the impedance.

Answer is C.

19. The peak current can be calculated as follows.

$$X_C = \frac{1}{\omega C}$$
$$= \frac{1}{\left(3000\ \dfrac{\text{rad}}{\text{s}}\right)(10 \times 10^{-6}\ \text{F})}$$
$$= 33.33\ \Omega \quad (33\ \Omega)$$
$$X_L = \omega L$$
$$= \left(3000\ \frac{\text{rad}}{\text{s}}\right)(20 \times 10^{-3}\ \text{H})$$
$$= 60\ \Omega$$
$$Z = R + j\,(X_L - X_C) = 51\ \Omega + j\,(60\ \Omega - 33\ \Omega)$$
$$= 51 + j27\ \Omega$$
$$= 57.7\ \Omega\angle 27.9° \quad (58\ \Omega\angle 27.9°)$$
$$i_P = \frac{v}{Z} = \frac{165\ \text{V}}{58\ \Omega\angle 27.9°}$$
$$= 2.8\ \text{A}\angle{-27.9°} \quad (2.8\ \text{A}\angle{-28°})$$

Note: In option (A), the resistance and reactances are simply added together. Options (B) and (D) demonstrate a misunderstanding of the relationship between peak voltage and rms voltage. In option (B), the 165 V peak voltage is divided by the square root of 2. In option (D), the peak voltage of 165 V is multiplied by the square root of 2.

Answer is C.

20. The resonant frequency is

$$f = \frac{1}{2\pi\sqrt{LC}}$$
$$= \frac{1}{2\pi\sqrt{(22 \times 10^{-6}\ \text{H})(100 \times 10^{-12}\ \text{F})}}$$
$$= 3.393 \times 10^6\ \text{Hz} \quad (3.4 \times 10^6\ \text{Hz})$$

Note: In option (A), the reciprocal is not taken. Option (B) is incorrect because the exponents are ignored. Option (D) is the radian (natural) frequency.

Answer is C.

21. The power delivered to the load is

$$P = VI\cos\theta = (440\ \text{V})(3.5\ \text{A})\cos 12°$$
$$= 1506\ \text{W} \quad (1510\ \text{W})$$

Note: Option (A) is the load resistance. In option (B), the sine is used instead of cosine, which yields the value for the apparent power. In option (D), the cosine factor was not used, giving the apparent power.

Answer is C.

22. The reactive power is

$$Q = VI\sin\theta$$
$$= (440\ \text{V})(3.5\ \text{A})\sin 12°$$
$$= 320\ \text{VAR}$$

Note: Option (B) does not use the correct units of volt-amperes-reactive (VAR), which are appropriate because power is not actually delivered. Instead, this erroneous calculation uses watts. Option (C) is the real power, not the reactive power. For option (D), the sine factor is not used.

Answer is A.

23. The quality factor for the circuit is

$$Q = \frac{f_0}{\text{BW}} = \frac{100\ \text{MHz}}{104.1\ \text{MHz} - 96.0\ \text{MHz}}$$
$$= 12.3 \quad (12)$$

Note: Option (A) is the bandwidth. Options (C) and (D) use the lower and upper halves of the bandwidth, respectively.

Answer is B.

24. The correct answer is the complex conjugate of the source Thevenin impedance. The complex conjugate is $75 - j34\ \Omega$.

Note: Option (B) is just the real part. Option (C) is not the complex conjugate. Option (D) is the Pythagorean theorem result.

Answer is A.

25. The transformer's output voltage is

$$V_{out} = V_{in} \frac{N_S}{N_P} = (117 \text{ V}) \left(\frac{660 \text{ turns}}{220 \text{ turns}} \right)$$

$$= 351 \text{ V} \quad (350 \text{ V})$$

Note: Option (A) would be true if the input voltage was direct current. For option (B), the number of primary turns and the number of secondary turns are inverted. Option (C) is the peak voltage.

Answer is D.

26. Transformers do not function with a DC input, so the correct answer is 0 V.

Note: All the other options fail to account for the fact that transformers do not function if the input is DC. Option (B) would be the right answer if the input was AC. Option (C) is the input voltage. Option (D) inverts the primary and secondary number of turns based on an AC input.

Answer is A.

27. The transformer turns ratio is

$$\frac{N_P}{N_S} = \sqrt{\frac{R_{source}}{R_{load}}} = \sqrt{\frac{330 \text{ }\Omega}{8 \text{ }\Omega}}$$

$$= 6.423 \quad (6.4 \text{ primary to secondary})$$

Note: Option (C) does not take the square root of the resistance ratios. In options (B) and (D), the secondary and primary values are inverted. In addition, in option (D) the square root is not taken.

Answer is A.

28. The turns ratio is

$$\frac{N_S}{N_P} = \frac{V_S}{V_P} = \frac{234 \text{ V}}{117 \text{ V}}$$

$$= 2.0$$

Therefore,

$$I_S = \frac{N_P}{N_S} I_P = \left(\frac{1}{2} \right) (2.2 \text{ A})$$

$$= 1.1 \text{ A}$$

Note: Option (B) is the turns ratio. Option (C) is the input current. The current is halved. The power in must equal the power out. In option (D), the voltage is doubled, but the current is halved. Again, the power in must equal the power out.

Answer is A.

29. The percentage efficiency is

$$\eta = \left(\frac{P_{out}}{P_{in}} \right) \times 100\%$$

$$= \frac{1600 \text{ W}}{(440 \text{ V}) (4.0 \text{ A})} \times 100\%$$

$$= 90.9\% \quad (90\%)$$

Note: Option (D) divides input power by output power, the inverse of the correct method. Option (C) would be correct if the problem described an ideal transformer, but this transformer has losses. Option (A) is the absolute efficiency, not the percentage efficiency.

Answer is B.

30. The parallel impedance can be calculated as follows.

$$B_C = \omega C = 2\pi f C$$

$$= 2\pi (60 \text{ Hz}) (20 \times 10^{-6} \text{ F})$$

$$= 0.007\,540 \text{ S} \quad (0.007\,5 \text{ S})$$

$$B_L = \frac{1}{\omega L} = \frac{1}{2\pi f L}$$

$$= \frac{1}{2\pi (60 \text{ Hz}) (20 \times 10^{-3} \text{ H})}$$

$$= 0.133 \text{ S}$$

$$G = \frac{1}{R} = \frac{1}{20 \text{ }\Omega} = 0.05 \text{ S}$$

$$Y = G + j(B_C - B_L)$$

$$= 0.05 \text{ S} + j(0.0075 \text{ S} - 0.133 \text{ S})$$

$$= 0.05 - j0.126 \text{ S}$$

$$= 0.136 \text{ S}\angle{-68.4°}$$

$$Z = \frac{1}{Y} = \frac{1}{0.136 \text{ S}\angle{-68.4°}}$$

$$= 7.35 \text{ }\Omega\angle{68.4°}$$

$$= 2.7 + j6.8 \text{ }\Omega$$

Note: In options (A) and (D), the conductances and susceptances are added without any imaginary component. Admittance and impedance are complex quantities. Also, in option (A), the admittance is not inverted to give impedance. In option (B), inductive susceptance is added instead of subtracted.

Answer is C.

3 Three-Phase Power and Electric Machines

Nomenclature

E_b	electromotive force	V
f	frequency	Hz
I	current	A
k	armature constant	–
l	length	m
L	inductance	H
n	rotational speed	rev/min
N	number of turns	–
p	number of poles	–
P	power	W
r	radius	m
R	resistance	Ω
s	slip	–
SR	speed regulation	%
t	time	s
T	torque	N·m
V	voltage	V
VR	voltage regulation	%
Z	impedance	Ω

Symbols

Φ	magnetic flux	Wb
θ	angle	degree
θ	phase angle difference	degree
ω	angular speed	rad/s

Subscripts

AC	alternating current
line-neutral	line to neutral
line-line	line to line
AN	line (phase) A to neutral
BN	line (phase) B to neutral
CN	line (phase) C to neutral
AB	line (phase) A to line (phase) B
BC	line (phase) B to line (phase) C
CA	line (phase) C to line (phase) A
A	line (phase) A
B	line (phase) B
C	line (phase) C

THREE-PHASE POWER

Three-phase systems use three wires to transmit power. (One configuration has a fourth wire for neutral or ground, but it does not carry much current.) Each of the three wires transmits full power. Three-phase motors are used for heavy industrial applications because of their greater efficiency. Most high voltage transmission lines have three main transmission cables. Some have an additional two wires directly attached to the top of the towers for lightning protection, but these do not carry the load. In such a five-wire configuration, the three power-carrying conductors are below the two lightning-protection wires. Three-phase transmission is three times as efficient as a single-phase system.

In contrast, single-phase power transmission uses two wires to transmit power. One wire sends the power to the load, and the other wire is the circuit return.

The voltage on each wire in a three-phase system is 120° phase shifted relative to the voltage on the other wires, as shown in the oscilloscope trace in Fig. 3.1. The phasor diagram for three-phase transmission is shown in Fig. 3.2.

Figure 3.1 Three-Phase Oscilloscope Trace

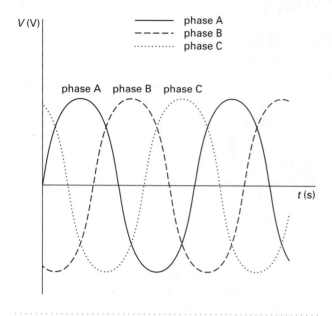

Figure 3.2 Three-Phase Phasor Diagram

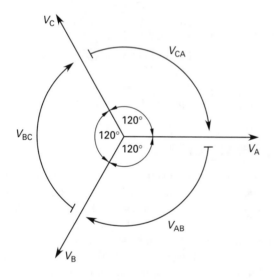

Figure 3.3 shows a simplified sketch of a three-phase generator. It has three coils located 120° apart. As the magnet rotates, it induces a sine wave in each coil that is 120° apart in phase.

The direction of rotation in three-phase motors depends on the phase sequence. This means a motor's direction can be reversed by interchanging two of the phase wires.

THREE-PHASE TRANSMISSION

It is possible to use six wires, two for each coil, making three single-phase systems. However, it works as well to use three wires in a three-phase system.

Figure 3.3 Simplified Three-Phase Generator

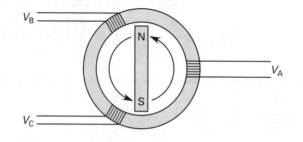

There are two types of three-phase power transmission: the delta configuration and the wye (Y-shaped) configuration.

Wye Configuration

The three-phase wye configuration has three lines (legs or phases) and a neutral (ground) wire. This configuration is shown in Fig. 3.4. Phase voltages are between legs (i.e., in the illustration, they are named V_{AB}, V_{BC}, and V_{CA}. Line-to-neutral voltages are V_{AN}, V_{BN}, and V_{CN}). *Line currents*, or *phase currents*, are the currents in lines A, B, and C.

If the effective alternating-current (AC) voltage between each leg and neutral is 120 V (household voltage), the voltage between legs is $(120 \text{ V})\sqrt{3} = 208$ V. The 120 V from each line to neutral is the common voltage available from wall outlets in North America. The line-to-neutral voltage is

$$V_{\text{line-neutral}} = \frac{V_{\text{line-line}}}{\sqrt{3}} = \frac{208 \text{ V}}{\sqrt{3}}$$
$$= 120 \text{ V}$$

Figure 3.4 Three-Phase Wye Configuration

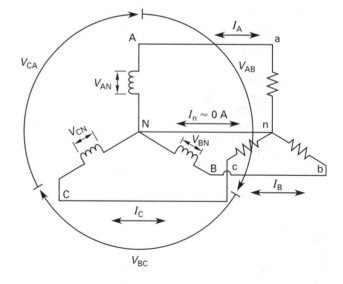

Line-to-line voltages and line-to-neutral voltages are 30° out of phase. For instance, the phase voltage V_{AB} leads the line-to-neutral voltage V_{AN} by 30°.

$$V_{AB} = \sqrt{3}V_{AN}\angle 30° \qquad 3.1$$

Or, the description could be that the line-to-neutral voltage lags the line-to-line voltage by 30°.

$$V_{AN} = \frac{V_{AB}}{\sqrt{3}} \angle -30° \qquad 3.2$$

Here, the V_{AN} voltage was used as reference to the V_{AB} voltage, but this relationship holds true for the other lines.

$$V_{BN} = V_{AN}\angle -120° \qquad 3.3$$
$$V_{CN} = V_{AN}\angle 120° \qquad 3.4$$

Example 3.1

Given a three-phase transmission system with 4400 V of line-to-line voltage, what is the line-to-neutral voltage?

Solution

$$\begin{aligned} V_{AN} &= \frac{V_{AB}}{\sqrt{3}} \angle -30° \\ &= \left(\frac{4400\text{ V}}{\sqrt{3}}\right) \angle -30° \\ &= 2540\text{ V}\angle -30° \end{aligned}$$

Example 3.2

A three-phase system has 220 V from line A to neutral. If the configuration looks like Fig. 3.4, what is its line-to-line V_{AB} voltage?

Solution

$$\begin{aligned} V_{AB} &= \sqrt{3}V_{AN}\angle 30° \\ &= \sqrt{3}(220\text{ V})\angle 30° \\ &= 381\text{ V}\angle 30° \end{aligned}$$

Delta Configuration

The three-phase delta configuration can be thought of as three separate generators operating 120° apart. Figure 3.5 shows the delta configuration. The generator is on the left and the load on the right. A typical light industrial line-to-line voltage is 208 V.

The line currents are out of phase with the line-to-line currents.

$$I_A = \sqrt{3}I_{AB}\angle -30° \qquad 3.5$$
$$I_{AB} = \frac{I_A}{\sqrt{3}} \angle 30° \qquad 3.6$$

Figure 3.5 Three-Phase Delta Configuration

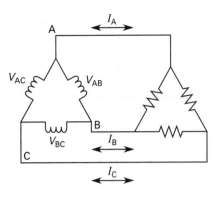

Example 3.3

A three-phase delta system has a load (line-to-line) current of 3.3 A. What are the line currents?

Solution

$$\begin{aligned} I_A &= \sqrt{3}I_{AB}\angle -30° \\ &= \sqrt{3}(3.3\text{ A})\angle -30° \\ &= 5.7\text{ A}\angle -30° \end{aligned}$$

Example 3.4

A three-phase delta system has line currents of 4.4 A. What is the line-to-line load current?

Solution

$$\begin{aligned} I_{AB} &= \frac{I_A}{\sqrt{3}} \angle 30° \\ &= \left(\frac{4.4\text{ A}}{\sqrt{3}}\right) \angle 30° \\ &= 2.5\text{ A}\angle 30° \end{aligned}$$

BALANCED LOADS

If the currents in each leg are identical in magnitude and 120° apart in phase, the three-phase system is said to be balanced.

Generators and loads can be either delta or wye. If a wye configuration is balanced, no current flows in the neutral wire. If the loads are not balanced between the three-phase lines, current will flow in the neutral wire. Often the neutral wire is a smaller gauge than the A, B, and C lines because it will be carrying much less current. It is desirable to keep the loads as balanced as reasonably possible to keep the neutral current as small as possible.

Three-Phase Power in a Balanced Load

Power in each line is calculated using the following equation. V_{line} is the line voltage, I_{line} is the line current, and $\cos\theta$ is the power factor. In systems in which V_{line} is the line-to-line voltage (rather than the line voltage) and I_{line} is the line current, $\cos\theta$ is usually the phase angle of the load impedance.

$$P_{\text{line}} = V_{\text{line}} I_{\text{line}} \cos\theta \qquad 3.7$$

Be sure to remember that $\cos\theta$ is the power factor (usually load impedance angle), and not the phase angle between the legs.

The line-to-neutral voltage is

$$V_{\text{line-neutral}} = \frac{V_{\text{line-line}}}{\sqrt{3}}$$

Given the line-to-line voltage, line current, and power factor, the power transferred is

$$P_{\text{line}} = \frac{V_{\text{line}} I_{\text{line}}}{\sqrt{3}} \cos\theta \qquad 3.8$$

Example 3.5

Given a 208 V line-to-neutral, three-phase system with a current of 2.0 A in each line and a power factor of 1.0, find the power delivered on each line and the total power delivered.

Solution

Power in each line is

$$\begin{aligned}
P_{\text{line}} &= V_{\text{line}} I_{\text{line}} \cos\theta \\
&= (208\text{ V})\,(2\text{ A})\cos 0° \\
&= 416\text{ W}
\end{aligned}$$

Total power is

$$\begin{aligned}
P_{\text{total}} &= 3P_{\text{line}} \\
&= (3)(416\text{ W}) \\
&= 1248\text{ W}
\end{aligned}$$

This is three times the power that could be delivered with a single-phase system operating with the same parameters.

Example 3.6

A three-phase system has a 440 V line-to-line voltage and a 4.2 A line current. The load impedance has a phase angle of 32°. Find the line power.

Solution

$$\begin{aligned}
P_{\text{line}} &= V_{\text{line}} I_{\text{line}} \cos\theta \\
&= \left(\frac{440\text{ V}}{\sqrt{3}}\right)(4.2\text{ A})\cos 32° \\
&= 905\text{ W}
\end{aligned}$$

ELECTRIC MACHINES

Electric machines are either motors or generators. *Motors* convert electrical energy into mechanical energy. *Generators* convert mechanical energy into electrical energy. Motors and generators are very similar; in fact, most motors will functions as generators and most generators will function as motors.

Generators and motors are both rotating machines. The *rotor* is the rotating part. The *stator* is the stationary part. The *armature* is the wire winding on the rotor. The armature carries the magnetizing current.

Motors depend on magnetic attraction and repulsion to create mechanical energy. The attraction and repulsion occur between the armature and the stator, giving the motor a rotating force. The magnetic field orientations change as the rotor rotates. This is accomplished in AC machines by the changing current. In DC machines, this function is handled by commutators, which are essentially switches controlled by the rotor position.

Generators use wires passing through a magnetic field to create a voltage and current. Magnetic fields are present in permanent magnets, and are also created by current flowing in wires. Permanent magnets can have their magnetization increased by electric currents flowing through magnet windings.

Speed and Voltage Regulation

Motors will slow down when they have a load. Their rotational speed in rpm, n, declines. The percentage of speed regulation, SR, is

$$\text{SR} = \left(\frac{n_{\text{no load}} - n_{\text{rated load}}}{n_{\text{rated load}}}\right) \times 100\% \qquad 3.9$$

Generator output voltage will decrease with a load. The percentage of voltage regulation, VR, is

$$\text{VR} = \left(\frac{V_{\text{no load}} - V_{\text{rated load}}}{V_{\text{rated load}}}\right) \times 100\% \qquad 3.10$$

This can be modeled with a Thevenin equivalent (see Ch. 1).

Example 3.7

An electric motor's unloaded speed is 1800 rpm. At the rated load, the speed drops to 1710 rpm. What is the percentage of speed regulation?

Solution

$$\begin{aligned}
\text{SR} &= \left(\frac{n_{\text{no load}} - n_{\text{rated load}}}{n_{\text{rated load}}}\right) \times 100\% \\
&= \left(\frac{1800\,\dfrac{\text{rev}}{\text{min}} - 1710\,\dfrac{\text{rev}}{\text{min}}}{1710\,\dfrac{\text{rev}}{\text{min}}}\right) \times 100\% \\
&= 5.26\%
\end{aligned}$$

Example 3.8

An electric generator has a no-load output of 451 V and a rated-load output of 439 V. What is its percentage of voltage regulation?

Solution

$$\text{VR} = \left(\frac{V_{\text{no load}} - V_{\text{rated load}}}{V_{\text{rated load}}} \right) \times 100\%$$

$$= \left(\frac{451 \text{ V} - 439 \text{ V}}{439 \text{ V}} \right) \times 100\%$$

$$= 2.73\%$$

Motor Types

There are three electric motor types: direct current (DC), synchronous, and induction. DC motors can work on either AC or DC depending on the configuration. Synchronous and induction motors run on AC only.

Induction motors are cheap and quite rugged. They are easily rewound and maintained, less sensitive to environmental changes than other motors, and do not require brushes or slip rings as do DC motors. The induction motor is generally a constant speed device, but relatively new *variable frequency drives* can vary the motor speed. The rotational speed of an induction motor is somewhat less than a submultiple of the driving line current frequency. The difference between the rotational speed and the submultiple of the driving frequency is called *slip*.

The torque-speed (*T*-*n*) graph of a typical induction motor driving a fan is shown in Fig. 3.6. The dotted line is the mechanical load line typical of a fan, and the solid line is the motor torque. The fan's mechanical loading typically increases as the square of the speed. If the speed slows from the operating point, the motor provides more torque. If the speed increases, the torque decreases, slowing the fan. These characteristics keep the induction motor speed relatively constant.

Figure 3.6 Induction Motor Operating Curve

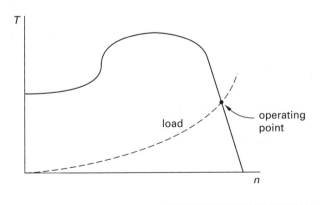

DC motors have excellent speed control and good torque down to zero rotational speed. Typical applications are electric drills and hand tools and vehicle motors, including locomotives. Their disadvantages are the need for slip rings and brushes, and the dangerous possibility of speed run away if not loaded. AC-to-DC supplies have been developed that convert AC to DC right at the motor rotor, eliminating the need for brushes and slip rings. Figure 3.7 shows the operating curve for a typical DC motor.

Figure 3.7 DC Motor Operating Curve

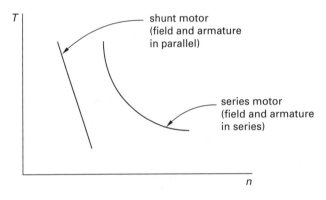

Synchronous motors rotate at a submultiple of the driving AC frequency. This makes them useful for clocks and timing devices. Synchronous machines are used as generators in large power plants that provide constant 60 Hz or 50 Hz frequency power. When load torque is increased, the rotor slips back a few degrees (fall back) but continues to run in synchronism. Of course, there is a practical limit to the load torque at which the motor is no longer able to run synchronously.

Faraday's Law

Faraday's law is the basis of electrical generators that turn mechanical energy into electrical energy. The law states that current and voltage are generated when a conductor passes through a magnetic field, as shown in Figs. 3.8 and 3.9.

The generated voltage (induced voltage) can be calculated using the following equation. *E* is the generated voltage, *N* is the number of coil turns, Φ is the magnetic flux in webers, θ is the angle between the magnetic flux and the direction of the conductor movement, and *t* is time in seconds.

$$E = -N \frac{\Delta \Phi}{\Delta t} \sin \theta \qquad 3.11$$

The faster the coil motion and the greater the flux, the greater the voltage. The voltage generated is maximized when the coil motion is perpendicular to the magnetic flux direction ($\sin \theta = 0$), and 0 V when the coil motion is parallel to the magnetic field direction.

Figure 3.8 Induced Voltage

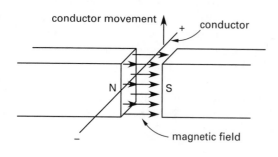

Figure 3.9 Induced Current

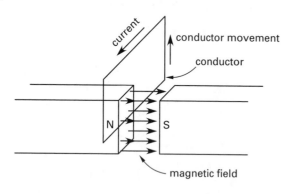

Right Hand Rule

Electrons in motion (electrical current) create a magnetic field. Figure 3.10 shows the end view of a magnetic field created by an electrical current. The direction of the magnetic field follows the right hand rule: if the right hand thumb points in the current direction, the fingers will point in the magnetic field direction.

Figure 3.10 Magnetic Fields Created by Current Flow (End View)

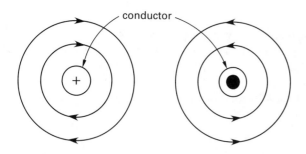

Current flowing into page. Right hand thumb points into page. Fingers point clockwise.

Current flowing out of page. Right hand thumb points out of page. Fingers point counterclockwise.

Magnetic Field Interaction

Motors and generators have permanent magnets on their stators, creating a stationary magnetic field, and have current-carrying conductors in their armatures, generating a moving magnetic field. Interaction between the armature magnetic field and the stator magnetic field makes generators and motors work.

A current-carrying conductor (armature) creates a magnetic field, which interacts with the stator magnetic field, as shown in Fig 3.11.

Figure 3.11 Interaction of Magnetic Fields in an Electric Motor

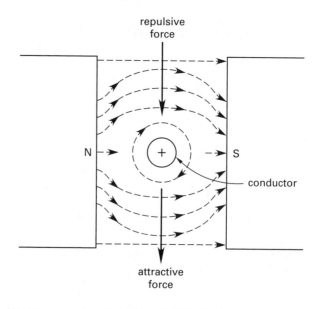

In an electric motor, this interaction between the magnetic fields drives the conducting wire (armature and rotor), pushing it with repulsive force and pulling it with attractive force. As it does so, this interaction converts electrical energy into mechanical energy.

The force on the wire can be calculated with the following equation. F is the mechanical force on the wire, B is the magnetic flux density, and θ is the angle between the wire movement and the magnetic field.

$$F = IB\sin\theta \qquad 3.12$$

A generator will supply power (electrical energy) if there is a load (resistor) on the conducting wire. This power requires an external force—mechanical energy input—as shown in Fig. 3.12. The external mechanical force is needed to overcome the electrical force produced by the interaction of the magnetic fields in the generator.

Commutation in DC Machines

A motor like the one in Fig. 3.11 would only give a single push as the wire passed the permanent magnet. A

Figure 3.12 *Interaction of Magnetic Fields in an Electric Generator*

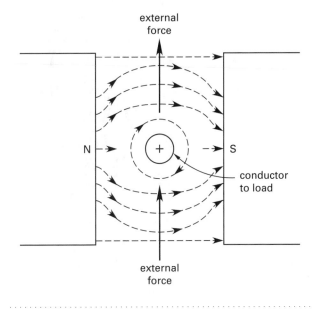

Figure 3.13 *Simple DC Electric Motor*

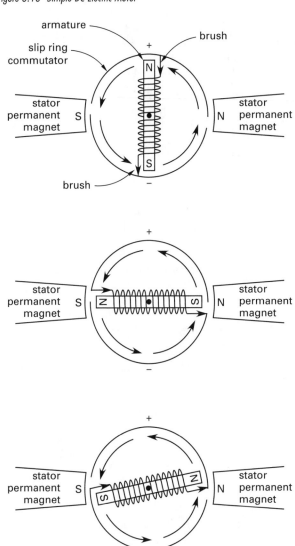

generator like the one in Fig. 3.12 would only produce a blip of energy as the conductor was forced past the permanent magnet. A means of reversing a magnetic field is needed to keep the motor running and the generator producing energy.

A commutator in a DC motor changes current direction and thus magnetic field direction. Figure 3.13 shows a simple DC motor. The rotor coil receives current through the copper slip rings. Spring-loaded carbon brushes push against the copper slip rings, allowing the current to pass through the armature coil. The current magnetizes the rotating armature. When the armature has rotated 180° the brushes contact the opposite slip ring, so the armature current direction is reversed, and so the armature magnetic field direction is reversed. This brush-slip ring combination is the commutator, a switch dependent on the rotor position.

In the top illustration in Fig. 3.13, the upper part of the armature is magnetized north and the lower part of the armature is magnetized south. The armature's north pole is attracted to the south pole of the stator's permanent magnet and repulsed from the north pole of the stator's permanent magnet. The armature south pole is attracted to the stator permanent magnet north pole and repulsed from the stator permanent magnet south pole. Under these forces, the armature rotates counterclockwise, as shown by the arrows.

In the middle illustration in Fig. 3.13, the armature has rotated so that its magnetic poles are at maximum attraction. The brushes are about to leave contact with

their copper slip rings and contact the opposite polarity slip ring.

In the bottom illustration in Fig. 3.13, the counterclockwise momentum of the rotor carries the brushes to the opposite polarity slip rings. The armature's magnetic polarity is reversed and the poles repel each other. The armature continues its counterclockwise rotation and the process repeats.

A popular middle school science project is to make a simple DC motor like Fig. 3.13. A wire-wound nail is the armature. The split copper ring is put on a cork. Two wire-wound nails become the stator magnets.

Figure 3.14 shows various methods of exciting the stator (field) magnets and the armature. The separately excited configuration will only work with both sources

AC or both sources DC, not with sources that are mixed AC and DC. The shunt-excited and series-connected configurations can work with DC sources, AC sources, or mixed AC-DC sources. As these configurations suggest, although this type of motor is referred to as a DC motor, it will work with either DC or AC.

Figure 3.14 Field and Armature Excitation

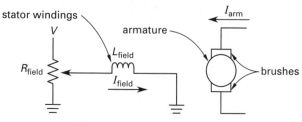

separately excited armature and field

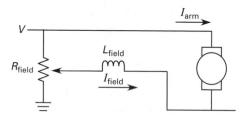

shunt-excited armature and field

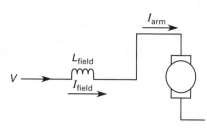

series-connected armature and field

Figure 3.15 Armature

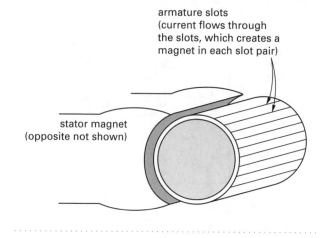

Figure 3.16 Rotor

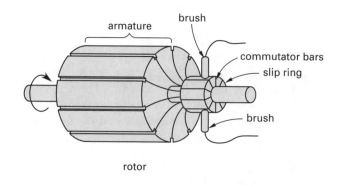

Figure 3.17 Armature End View

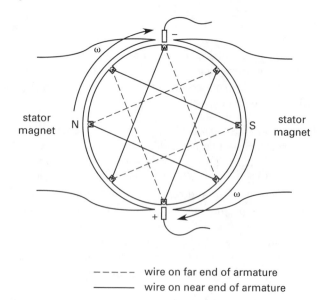

Actual motor armatures (rotors) have many magnets, as shown in Fig. 3.15. Figure 3.16 shows the rotor with six windings (three pairs).

Figure 3.17 is a view of the armature showing how it is wound. In practice, many wires pass through the slots. The wires in each slot create an individual magnetic field, thus there are as many different magnets as one half the number of slots. Only one wire is shown for simplicity. The wires are typically insulated with an enamel coating.

The following equation can be used to calculate a motor's armature torque in newton-meters. N is the number of active conductors, r is armature radius in meters,

l is armature length in meters, B is the magnetic field density in teslas, and I is the armature current in amperes.

$$T = 0.5NrlBI \qquad 3.13$$

The output power in watts is the product of the armature's torque multiplied by its angular rotational speed in radians per second, ω, as shown in the following equation.

$$P = T\omega \qquad 3.14$$

Example 3.9

A DC motor has 24 slots with 16 slots active, 0.8 T magnetic field density, length of 1.0 m, 0.35 m radius, and an armature current of 55 A. It is operating at 1500 rpm. What is the armature torque and power? Give the power in watts and horsepower. There are 746 W in 1 hp.

Solution

$$\begin{aligned}
T &= 0.5NrlBI \\
&= (0.5)\,(16)\,(0.35\text{ m})\,(1.0\text{ m})\,(0.8\text{ T})\,(55\text{ A}) \\
&= 123.2\text{ N·m} \quad (120\text{ N·m}) \\
P &= T\omega \\
&= (120\text{ N·m})\left(\frac{2\pi\left(1500\ \dfrac{\text{rev}}{\text{min}}\right)}{60\ \dfrac{\text{s}}{\text{min}}}\right) \\
&= 18\,850\text{ W} \\
P_{\text{hp}} &= (18\,850\text{ W})\left(\frac{1\text{ hp}}{746\text{ W}}\right) \\
&= 25.27\text{ hp}
\end{aligned}$$

Back emf

A motor actually is both a motor and a generator. The armature rotation causes the generator effect, producing a back electromotive force (emf) that opposes the applied voltage. The generated back emf, E_b, can be calculated with the following equation. Φ is the magnetic flux, ω_m is the rotational speed in radians per second, and k is the armature constant.

$$E_b = k\Phi\omega_m \qquad 3.15$$

The power generated by the motor is the product of the back emf and the armature current.

$$P = E_b I_{\text{arm}} = k\Phi\omega_m I_{\text{arm}} \qquad 3.16$$

The equivalent circuit is shown in Fig. 3.18.

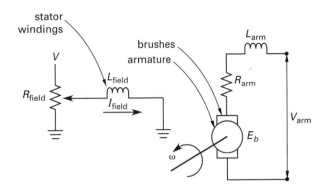

Figure 3.18 Separately Excited DC Motor Equivalent Circuit

When the rotational velocity is 0 rpm (at start-up), the back emf is 0 V. The armature start-up current is

$$I_{\text{arm}} = \frac{V_{\text{arm}}}{Z_{\text{arm}}} = \frac{V_{\text{arm}}}{R_{\text{arm}}} \qquad 3.17$$

If the rotor is locked and cannot rotate, the current may exceed the motor power limitations and the motor may be damaged. Insulation damage due to overheating is a common type of damage. Motors are usually protected by slow-blow fuses or circuit breakers. The slow-blow fuses are designed to withstand a significant overload (start-up current) for several seconds before opening. If the rotor is locked, the fuse will blow (open) in 3-5 s. When the armature is running at proper velocity, the armature current is

$$I_{\text{arm}} = \frac{V_{\text{arm}} - E_b}{Z_{\text{arm}}} = \frac{V_{\text{arm}} - E_b}{R_{\text{arm}}} \qquad 3.18$$

A magnetization curve is shown in Fig. 3.19. It is found by rotating the machine at a constant speed, varying the field current and with it the stator magnetism. There is no load, so there is no output voltage drop. The output voltage is the back emf.

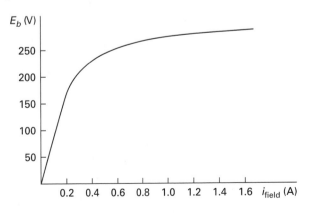

Figure 3.19 Generator Output

Example 3.10

An electric motor has an armature constant of 150, a flux of 22 mWb, an armature current of 28 A, and a rotational velocity of 1200 rpm. What is the power output in watts and horsepower?

Solution

$$\omega = 2\pi n$$

$$= \left(2\pi \frac{\text{rad}}{\text{rev}}\right) \left(\frac{1200 \frac{\text{rev}}{\text{min}}}{60 \frac{\text{s}}{\text{min}}}\right)$$

$$= 126 \text{ rad/s}$$

$$P = E_b I_{\text{arm}}$$

$$= k\Phi\omega I_{\text{arm}}$$

$$= (150)\left(22 \times 10^{-3} \text{ Wb}\right)\left(126 \frac{\text{rad}}{\text{s}}\right)(28 \text{ A})$$

$$= 11\,642 \text{ W} \quad (11.6 \text{ kW})$$

$$P_{\text{hp}} = (11.6 \text{ kW})\left(\frac{1 \text{ hp}}{0.746 \text{ kW}}\right)$$

$$= 15.55 \text{ hp}$$

Example 3.11

A separately excited generator has the magnetization curve shown in Fig. 3.19. It is running at 1500 rpm with a field current of 0.2 A. The load is 90 Ω and the armature resistance is 0.7 Ω. What is the load voltage and current?

Solution

From the magnetization curve, a field current of 0.2 A gives an output of 150 V at 1800 rpm. The generator is operating at 1500 rpm. The back emf at 1500 rpm is

$$E_b = (150 \text{ V})\left(\frac{1500 \frac{\text{rev}}{\text{min}}}{1800 \frac{\text{rev}}{\text{min}}}\right)$$

$$= 125 \text{ V}$$

$$I_{\text{arm}} = \frac{E_b}{R_{\text{arm}} + R_{\text{load}}}$$

$$= \frac{125 \text{ V}}{0.7 \text{ Ω} + 90 \text{ Ω}}$$

$$= 1.38 \text{ A}$$

By the voltage divider rule,

$$V_{\text{load}} = E_b \left(\frac{R_{\text{load}}}{R_{\text{load}} + R_{\text{arm}}}\right)$$

$$= (125 \text{ V})\left(\frac{90 \text{ Ω}}{90 \text{ Ω} + 0.7 \text{ Ω}}\right)$$

$$= 124 \text{ V}$$

Induction Machines and Synchronous Machines

Induction machines are AC machines and do not need brushes, slip rings, or commutators. AC currents in the stator windings of a three-phase motor change as the supply voltage cycles, so the stator magnetic fields rotate with the AC cycle. The stator and the armature are a transformer. The rotating magnetic field is transformer coupled to the armature, and pulls the rotor along with the magnetic field. An induction motor rotates at a speed less than the stator rotating magnetic field.

An induction motor transformer couples the stator magnetic field to the rotor. Figure 3.20 shows a magnet moving across a conductive metal plate. The moving magnetic field induces an electric current in the metal plate. This may be thought of as a transformer with a shorted secondary.

Figure 3.20 Magnetic Induction

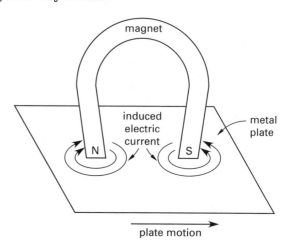

An induction motor has an armature similar to the one in Fig. 3.21. The rotor is called a *squirrel cage rotor* because it looks similar to exercise wheels for small rodents. The rotor is usually cast aluminum wrapped around a ferrous core. The rotating stator magnetic field induces electric fields in the aluminum conductors, similar to the method shown in Fig. 3.20.

For this induction to happen, there must be a speed difference between the rotor and rotating stator magnetic field. The speed difference creates the changing magnetic field necessary for transformer action. Therefore the induction motor's rotational speed is always less than the rotating stator magnetic field speed. This difference is called slip. Slip, s, can be calculated with two forms of an equation in which n_s is the synchronous speed and n is the actual speed.

$$s = \frac{n_s - n}{n_s} \qquad 3.19$$

Figure 3.21 *Squirrel Cage Armature*

Figure 3.21 *Squirrel Cage Armature*

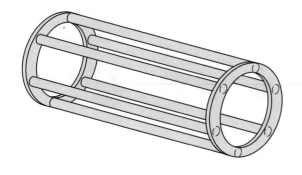

Figure 3.22 *Six-Pole Machine*

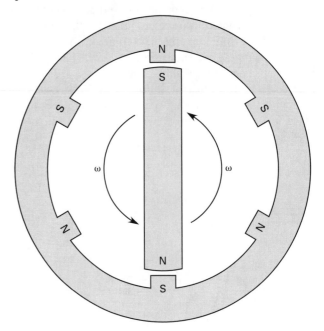

$$n = n_s (1 - s) \qquad 3.20$$

A synchronous motor's rotor stays in phase (aligned) with the stator magnetic field. A DC current is applied to the rotor coils through slip rings to create the magnetic field that aligns the rotor to the stator rotating field.

Example 3.12

An induction motor operates at a speed of 1100 rpm under full load, and its synchronous speed is 1200 rpm. What is the slip amount?

Solution

$$
\begin{aligned}
s &= \frac{n_s - n}{n_s} \\
&= \frac{1200 \ \dfrac{\text{rev}}{\text{min}} - 1100 \ \dfrac{\text{rev}}{\text{min}}}{1200 \ \dfrac{\text{rev}}{\text{min}}} \\
&= 0.083
\end{aligned}
$$

Poles and Motor Speed

Thus far, only motors with two stator poles (one pole pair) have been discussed. Usually, many pole pairs are used in electrical machines. Figure 3.22 shows a six-pole stator. The protruding poles are called *salient poles*. The synchronous frequency is given by the following equation, in which f is the generated frequency in hertz, p is the number of poles and $p/2$ is the number of pole pairs (north-south), and n_s is the synchronous speed in revolutions per minute.

$$f = \frac{p}{2} \left(\frac{n_s}{60 \ \dfrac{\text{s}}{\text{min}}} \right) \qquad 3.21$$

Example 3.13

A 12-pole synchronous generator outputs 60 Hz power. What is the necessary rotational speed?

Solution

$$
\begin{aligned}
f &= \frac{p}{2} \left(\frac{n_s}{60 \ \dfrac{\text{s}}{\text{min}}} \right) \\
n_s &= \frac{2}{p} \left(60 \ \frac{\text{s}}{\text{min}} \right) f \\
&= \left(\frac{2}{12} \right) \left(60 \ \frac{\text{s}}{\text{min}} \right) \left(60 \ \frac{\text{cycles}}{\text{s}} \right) \\
&= \left(600 \ \frac{\text{rev}}{\text{min}} \right) \left(\frac{1 \ \text{min}}{60 \ \text{s}} \right) \\
&= 10 \ \text{rev/s}
\end{aligned}
$$

Example 3.14

A hydroelectric turbine generator must output 60 Hz power. The generator rotates at 180 rpm. What is the required number of poles?

Solution

$$f = \frac{p}{2} \left(\frac{n_s}{60 \ \dfrac{\text{s}}{\text{min}}} \right)$$

$$
\begin{aligned}
p &= \frac{(2) \left(60 \ \dfrac{\text{s}}{\text{min}} \right) f}{n} \\
&= \frac{(2) \left(60 \ \dfrac{\text{s}}{\text{min}} \right) (60 \ \text{Hz})}{180 \ \dfrac{\text{rev}}{\text{min}}} \\
&= 40 \ \text{poles} \quad \text{(20 north-south pole pairs)}
\end{aligned}
$$

Start-Up

Single phase synchronous motors and some induction motors have zero torque before start-up (0 rpm), preventing them from starting. The common solution is an auxiliary winding, as shown in Fig. 3.23. The winding has an electrical phase shift provided by a resistor, as shown in the top illustration, or a capacitor, as shown in the bottom illustration. When the motor nears operating speed, the centrifugal switch opens and the motor runs normally.

Figure 3.23 Auxiliary Winding Solutions for Start-Up

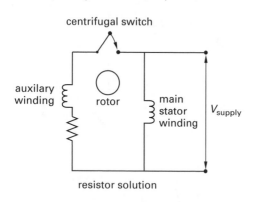

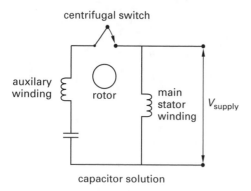

Variable-Frequency Drive Motors

Electric motor speed and torque can be more closely controlled if the motor-driving frequency is variable. Variable-frequency motor controllers are made by converting the line AC voltage into a DC voltage, then inverting the DC voltage to a variable frequency AC voltage driving the motor, as shown in Fig. 3.24. Such controllers are usually applied to induction motors, but they can also be applied to synchronous motors.

The variable DC voltage is controlled by silicon-controlled rectifiers (SCR). This DC voltage often has a lot of *ripple*; that is, AC voltage riding on the DC voltage as shown in Fig. 3.25. Ripple can be reduced by techniques discussed in Ch. 4, but ripple does not significantly affect most electric motors.

Figure 3.24 Variable-Frequency Drive Motor

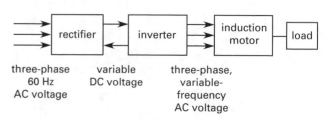

Figure 3.25 Ripple

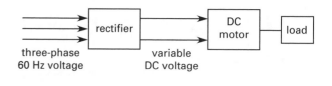

Variable-Voltage Drive Motors

A DC motor can be driven directly by a variable voltage, as shown in Fig. 3.26.

Figure 3.26 Variable-Voltage Drive Motor

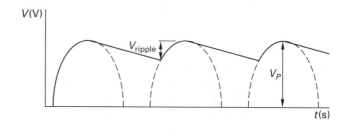

Brushless DC Motors

The brushless DC motor is not actually a DC motor, but a motor driven by a DC voltage that has been converted to a variable-frequency AC voltage, like the motor shown in Fig. 3.27. It operates much as a shunt DC motor with constant field current.

Some advantages to this type of motor are there is no brush wear and no spark (electrical arc), so the unit can be sealed to operate under liquids and in explosive environments.

Figure 3.27 Brushless DC Motor

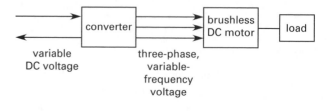

Stepping Motors

Stepping motors rotate in distinct angular steps. They can be driven in either direction or held in an angular position, or they can rotate at varying speeds. They are driven by logic signals and easily interface to digital computers. The angular position is controlled by current direction and by whether the current is on or off. Typical uses for these motors are printers and magnetic-disk drive head positioning. Typical stepping motors have 200 or more angular positions, with a typical change of 1.8° between each position. Figure 3.28 shows the operation of a four-position (four-pole) stepping motor. The plus and minus signs denote current direction.

Figure 3.28   Two-Phase, Four-Pole Stepping Motor

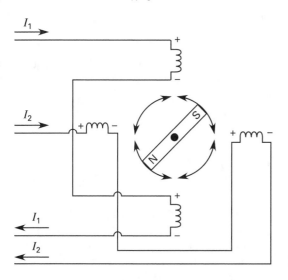

The rotational position will be 0°, 90°, 180°, and 270° if only one current flows at a time, as shown in Table 3.1.

Intermediate steps can be obtained by exciting both coils, as shown in Table 3.2.

Combining the two sequences gives eight positions, as shown in Table 3.3. Figure 3.29 shows the current waveforms.

Digital circuits make it easy to generate current sequences such as the sequence shown in Fig. 3.29. These digital circuits drive large power transistors, since the integrated circuits usually used for timing do not have high current outputs.

Single-Phase Induction Motors

For an electric motor to rotate, the magnetic field must rotate. This is easily accomplished with three-phase voltages. However, many home appliances only have single-phase power. In order to allow a single-phase motor to operate, an induction motor can be made with

Table 3.1   Single-Phase, Full-Step Sequence

θ	I_1	I_2
0°	+	0
90°	0	+
180°	−	0
270°	0	−
0°	+	0

Table 3.2   Two-Phase, Full-Step Sequence

θ	I_1	I_2
45°	+	+
135°	−	+
225°	−	−
315°	+	−
45°	+	+

Table 3.3   Half-Step Sequence

θ	I_1	I_2
0°	+	0
45°	+	+
90°	0	+
135°	−	+
180°	−	0
225°	−	−
270°	0	−
315°	+	−
0°	+	0

Figure 3.29   Half-Step Current Sequence

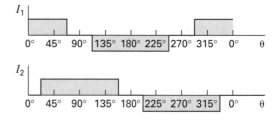

stator coils 90° apart, as shown in Fig. 3.30. This arrangement, called quadrature stator winding, creates a rotating magnetic field.

Figure 3.30   Single-Phase Induction Motor with Quadrature Stator Winding

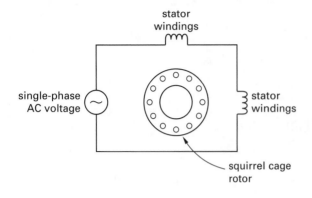

Universal Motor

A DC motor can be made to run on single-phase AC or DC by connecting the rotor and stator in series as shown in Fig. 3.31. The armature and field currents (and their magnetic fields) are then in phase.

Figure 3.31   Universal Motor

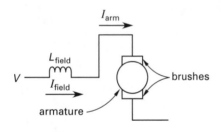

PRACTICE PROBLEMS

1. A three-phase power system has a 220 V line-to-line voltage. What is most nearly the line-to-neutral voltage?

    (A) 130 V
    (B) 130 V$\angle -30°$
    (C) 380 V$\angle 30°$
    (D) 380 V$\angle -30°$

2. A three-phase delta system has a line-to-line load current of 4.5 A. What are most nearly the line currents?

    (A) 2.6 A
    (B) 2.6 A$\angle 30°$
    (C) 7.8 A
    (D) 7.8 A$\angle -30°$

3. A three-phase delta system has a 2400 V line-to-line voltage and a 15 A current. The load impedance angle is 21°. Most nearly how much power is delivered in each line?

    (A) 19 kW
    (B) 34 kW
    (C) 36 kW
    (D) 58 kW

4. An electric motor's unloaded speed is 2400 rpm. The percentage speed regulation is 6.0%. What is most nearly the loaded speed?

    (A) 340 rpm
    (B) 400 rpm
    (C) 2300 rpm
    (D) 2600 rpm

5. A generator's no-load output is 2451 V, and its output at rated load is 2099 V. What is most nearly the percentage voltage regulation?

    (A) 0.17%
    (B) 14%
    (C) 17%
    (D) 350%

6. A DC motor has 36 slots with 24 slots active, a 0.9 T magnetic field density, a length of 1.2 m, a 0.40 m radius, and an armature current of 65 A. It is operating at 1600 rpm. What is most nearly the armature torque in newtons and the power in watts?

    (A) $T = 340$ N·m, $P = 9000$ W
    (B) $T = 340$ N·m, $P = 540\,000$ W
    (C) $T = 510$ N·m, $P = 13\,600$ W
    (D) $T = 680$ N·m, $P = 18\,000$ W

7. An electric motor has an armature constant of 140, a flux of 25 mWb, an armature current of 30 A, and a rotational speed of 1600 rpm. What is most nearly the power output in watts?

    (A) 2800 W
    (B) 17 600 W
    (C) 168 000 W
    (D) 17 600 000 W

8. A generator has the magnetization curve shown in the following illustration. The generator is running at 1200 rpm with a field current of 0.2 A. The load is 80 Ω and the armature resistance is 8.0 Ω. What is most nearly the load current and voltage?

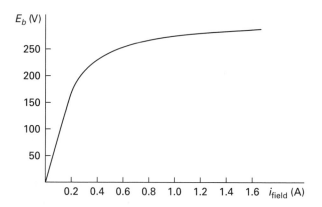

(A) 1.14 A, 90.9 V
(B) 1.25 A, 90.9 V
(C) 1.25 A, 100 V
(D) 1.70 A, 148 V

9. An induction motor operates at a speed of 1050 rpm under full load, and its synchronous speed is 1200 rpm. What is most nearly the slip amount?

(A) −0.13
(B) 0.13
(C) 0.14
(D) 13

10. A 20-pole synchronous generator is rotating at 300 rpm. What is the output frequency?

(A) 50 Hz
(B) 100 Hz
(C) 3000 Hz
(D) 6000 Hz

SOLUTIONS

1. The line-to-neutral voltage is

$$V_{\text{line-neutral}} = \frac{V_{\text{line-line}}}{\sqrt{3}} \angle -30°$$
$$= \frac{220 \text{ V}}{\sqrt{3}} \angle -30°$$
$$= 127 \text{ V} \angle -30° \quad (130 \text{ V} \angle -30°)$$

Note: Option (A) has the correct amplitude, but it does not include phase in the equation. In options (C) and (D), the line-to-line voltage equations are used.

Answer is B.

2. The line currents are

$$I_{\text{A}} = \sqrt{3}I_{\text{AB}}$$
$$= \sqrt{3}\,(4.5 \text{ A} \angle -30°)$$
$$= 7.794 \text{ A} \angle -30° \quad (7.8 \text{ A} \angle -30°)$$

Note: Options (A) and (C) do not include phase angle. Option (B) confuses line-to-line currents with line currents.

Answer is D.

3. The power delivered in each line is

$$P = V_{\text{line}} I_{\text{line}} \cos\theta$$
$$= (2400 \text{ V})\,(15 \text{ A}) \cos 21°$$
$$= 33\,609 \text{ W} \quad (34 \text{ kW})$$

Note: In options (A) and (D), the square root of three is mistakenly inserted. Option (C) does not use the phase angle.

Answer is B.

4. The loaded speed can be calculated as follows.

$$\text{SR} = \left(\frac{n_{\text{no load}} - n_{\text{rated load}}}{n_{\text{rated load}}}\right) \times 100\%$$

Therefore,

$$n_{\text{rated load}} = \frac{n_{\text{no load}}}{\dfrac{\text{SR}}{100\%} + 1} = \frac{2400 \dfrac{\text{rev}}{\text{min}}}{\dfrac{6.0\%}{100\%} + 1}$$
$$= 2264.15 \text{ rpm} \quad (2300 \text{ rpm})$$

Note: In option (A), the regulation percentage is not divided by 100%. In option (B), the wrong equation is used, the regulation percentage is not divided by 100%, and the unloaded speed was divided by 6. Option (D) also uses the wrong equation.

Answer is C.

5. The percentage voltage regulation can be calculated as follows.

$$\text{VR} = \left(\frac{V_{\text{no load}} - V_{\text{rated load}}}{V_{\text{rated load}}}\right) \times 100\%$$
$$= \left(\frac{2451 \text{ V} - 2099 \text{ V}}{2099 \text{ V}}\right) \times 100\%$$
$$= 16.8\% \quad (17\%)$$

Note: Option (A) is not multiplied by 100%. Option (B) uses the no-load voltage in the denominator. Option (D) is merely the difference between the no-load and rated-load voltages.

Answer is C.

6. The torque and power can be calculated as follows.

$$T = 0.5NrlBI$$
$$= (0.5)(24)(0.40 \text{ m})(1.2 \text{ m})(0.9 \text{ T})(65 \text{ A})$$
$$= 336.96 \text{ N·m} \quad (340 \text{ N·m})$$

$$P = T\omega$$
$$= (340 \text{ N·m})\left(\frac{1600 \frac{\text{rev}}{\text{min}}}{60 \frac{\text{s}}{\text{min}}}\right)$$
$$= 9067 \text{ W} \quad (9000 \text{ W})$$

Note: In option (B), the power is not divided by 60 s. In option (C), 36 slots were used instead of 24. Option (D) does not use the 0.5 multiplier.

Answer is A.

7. The power output can be calculated with the following equations.

$$\omega = \left(\frac{2\pi}{60 \frac{\text{s}}{\text{min}}}\right) n$$
$$= \left(\frac{2\pi}{60 \frac{\text{s}}{\text{min}}}\right)(1600 \text{ rpm})$$
$$= 167.55 \text{ rad/s} \quad (167.6 \text{ rad/s})$$

$$P = E_b I_{\text{arm}} = k\Phi\omega I_{\text{arm}}$$
$$= (140)(25 \times 10^{-3} \text{ Wb})\left(167.6 \frac{\text{rad}}{\text{s}}\right)(30 \text{ A})$$
$$= 17\,598 \text{ W} \quad (17\,600 \text{ W})$$

Note: Options (A) does not convert rpm to radians per second. Option (C) fails to use 2π in the numerator. Option (D) does not use the magnetic field intensity exponent properly.

Answer is B.

8. From the illustration, a field current of 0.2 A will give an output of 150 V at 1800 rpm. The generator is operating at 1200 rpm. E_b at 1200 rpm is

$$E_b = (150 \text{ V})\left(\frac{1200 \text{ rpm}}{1800 \text{ rpm}}\right)$$
$$= 100 \text{ V}$$

$$I_{\text{arm}} = \frac{E_b}{R_{\text{arm}} + R_{\text{load}}} = \frac{100 \text{ V}}{8.0 \ \Omega + 80 \ \Omega}$$
$$= 1.136 \text{ A} \quad (1.14 \text{ A})$$

By the voltage divider rule,

$$V_{\text{load}} = E_b \left(\frac{R_{\text{load}}}{R_{\text{load}} + R_{\text{arm}}}\right)$$
$$= (100 \text{ V})\left(\frac{80 \ \Omega}{80 \ \Omega + 8.0 \ \Omega}\right)$$
$$= 90.9 \text{ V}$$

Note: In option (B), the armature resistance is neglected in the current calculation. Option (C) neglects the armature resistance in the current and voltage calculations. Option (D) does not adjust the voltage from 1800 rpm to 1200 rpm.

Answer is A.

9. The slip amount is

$$s = \frac{n_s - n}{n_s}$$
$$= \frac{1200 \frac{\text{rev}}{\text{min}} - 1050 \frac{\text{rev}}{\text{min}}}{1200 \frac{\text{rev}}{\text{min}}}$$
$$= 0.125 \quad (0.13)$$

Note: In option (A), the unloaded speed is subtracted from the loaded speed, giving a negative value. Option (C) incorrectly uses the loaded speed as the divisor. Option (D) provides the percentage, which does not answer the problem.

Answer is B.

10. A 12-pole synchronous generator outputs 50 Hz power. The output frequency of a 20-pole synchronous generator is

$$f = \left(\frac{p}{2}\right)\left(\frac{n}{60 \frac{\text{s}}{\text{min}}}\right)$$
$$= \left(\frac{20}{2}\right)\left(\frac{300 \frac{\text{rev}}{\text{min}}}{60 \frac{\text{s}}{\text{min}}}\right)$$
$$= 50 \text{ Hz}$$

Note: In option (B), the number of poles is not divided by two to obtain the number of pole pairs (north-south poles). Options (C) and (D) do not use 60 s/min in the denominator to convert between rpm and herz. In addition, option (D) does not calculate the number of pole pairs.

Answer is A.

4 Operational Amplifiers and Diodes

Nomenclature

A	gain	–
I	current	A
R	resistance	Ω
V	voltage	V

Subscripts

DC	direct current
fb	feedback
in	input
in+	input to non-inverting input
in−	input to inverting input
n	total number of inputs
out	output
P	peak
PP	peak to peak
V	voltage

OPERATIONAL AMPLIFIERS

An *operational amplifier*, commonly called an op amp, is a high-gain differential amplifier. Typical gains are 100 000 V—too high a gain to be useful because these high gains result in circuit instability. External components tame the op amp and make it into a useful circuit. The external components are usually resistors and capacitors.

An op amp has one output and two differential inputs. The non-inverting input voltage is represented by V_{in+} and the inverting input voltage by V_{in-}. The output is proportional to the voltage difference between the two input terminals ($V_{in+} - V_{in-}$). Figure 4.1 shows the graphic symbol for an op amp, which also serves as a basic schematic diagram. The diagram does not show the op amp's internal workings. Also not shown are two power supply terminals.

Figure 4.1 Op Amp Symbol

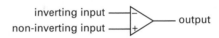

All modern op amps are integrated circuits, because an op amp made of discrete parts would be too unstable to be practical.

Op Amp Uses

Op amps are analog amplifiers that can perform mathematical operations. Some of the possible uses of op amps are

- summer: adds and subtracts analog voltages
- inverter: changes the phase of the signal by 180°
- buffer amplifier: provides a high input impedance that draws minimum current from the signal source and a low impedance output
- analog-signal integrator
- analog-signal differentiator
- high-pass filter: allows high frequencies to pass, but not low
- low-pass filter: allows low frequencies to pass, but not high
- band-pass filter: passes only a selected band of frequencies
- band-reject filter: rejects a certain frequency (e.g., rejects 60 Hz interference)
- voltage comparator: compares two voltages and the output indicates which is greater

INVERTING AMPLIFIERS

An op amp can be used to make an inverting amplifier, as shown in Fig. 4.2. The inverting amplifier components include an input resistor with resistance R_{in}, a feedback resistor with a resistance R_{fb}, and a virtual ground.

Figure 4.2 Inverting Amplifier

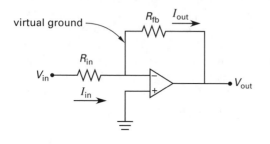

Virtual Ground

In an op amp, the noninverting input is grounded. The op amp gain is extremely high, so the voltage difference between inverting and noninverting inputs is essentially 0 V. An ideal op amp has infinite gain, and its ideal inputs do not draw current. The ideal op amp voltage gain, A_V, is calculated in the following equation, in which V_{out} is the output voltage and $V_{in+} - V_{in-}$ is the voltage difference between the noninverting input and the inverting input.

$$A_V = \frac{V_{out}}{V_{in+} - V_{in-}} \approx \infty \qquad 4.1$$

This means a tiny voltage difference at the input results in a large output change. Even small amounts of input noise can cause problems, possibly making an op amp difficult to control. But this high gain also leads to some interesting applications.

Since the gain is infinite,

$$V_{in+} - V_{in-} = \frac{V_{out}}{A_V} = \frac{V_{out}}{\infty}$$
$$\approx 0 \text{ V} \qquad 4.2$$

The voltage difference between the inverting and non-inverting inputs ($V_{in+} - V_{in-}$) is very small, essentially zero. In Fig. 4.2, the noninverting input is connected to ground. Since the voltage difference between the inputs is essentially zero, the junction of the two resistors and the inverting input is called a *virtual ground*. That point is at 0 V, but it is not actually grounded. The virtual ground makes analysis straightforward.

Inverting Amplifier Gain

In an inverting amplifier like the one in Fig. 4.2, with the right end of the input resistor at V_{in-} at virtual ground (0 V), and the signal input current is

$$I_{in} = \frac{V_{in} - V_{in-}}{R_{in}} = \frac{V_{in}}{R_{in}} \qquad 4.3$$

No current goes into the inverting input, so all the current must go out through the feedback resistor. Therefore, because the left end of the feedback resistor is at $V_{in-} = 0$ V, a virtual ground, so the input current equals the output current.

$$I_{out} = \frac{0 \text{ V} - V_{out}}{R_{fb}}$$
$$= \frac{-V_{out}}{R_{fb}}$$
$$= I_{in} \qquad 4.4$$
$$\frac{\text{signal input}}{R_{in}} = \frac{-\text{signal output}}{R_{fb}} \qquad 4.5$$

The voltage gain is

$$A_V = \frac{\text{signal output}}{\text{signal input}}$$
$$= -\frac{R_{fb}}{R_{in}}$$
$$= \frac{V_{out}}{V_{in}} \qquad 4.6$$

Note the circuit gain is independent of op amp gain.

Example 4.1

An inverting amplifier like the one in Fig. 4.2 has a feedback resistance of 3.0 kΩ and an input resistance of 1.2 kΩ. What is the circuit gain?

Solution

$$A = -\frac{R_{fb}}{R_{in}} = -\frac{3.0 \text{ k}\Omega}{1.2 \text{ k}\Omega}$$
$$= -2.5$$

SUMMING AMPLIFIERS

The inverting op-amp circuit can be modified to sum analog voltages.

The current I_{in} into the summing junction (virtual ground) is the sum of all input currents.

$$I_{in} = I_{in-1} + I_{in-2} + I_{in-3} \ldots I_{in-n}$$
$$= \frac{V_1}{R_{in-1}} + \frac{V_2}{R_{in-2}} + \frac{V_3}{R_{in-3}} + \ldots + \frac{V_n}{R_{in-n}}$$

Figure 4.3 Summing Amplifier

Figure 4.3 Summing Amplifier

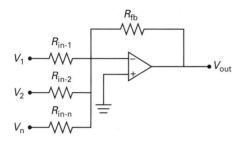

The input current is equal to the output current i_{out}.

$$I_{\text{out}} = \frac{0\text{ V} - V_{\text{out}}}{R_{\text{fb}}} = \frac{-V_{\text{out}}}{R_{\text{fb}}} = I_{\text{in}}$$

$$= \frac{V_1}{R_{\text{in}-1}} + \frac{V_2}{R_{\text{in}-2}} + \frac{V_3}{R_{\text{in}-3}} + \ldots + \frac{V_n}{R_{\text{in}-n}}$$

The voltage output is

$$V_{\text{out}} = \frac{R_{\text{fb}}}{I_{\text{in}-1} + I_{\text{in}-2} + I_{\text{in}-3} + \ldots + I_{\text{in}-n}}$$

$$= \frac{R_{\text{fb}}}{\dfrac{V_1}{R_{\text{in}-1}} + \dfrac{V_2}{R_{\text{in}-2}} + \dfrac{V_3}{R_{\text{in}-3}} + \ldots + \dfrac{V_n}{R_{\text{in}-n}}}$$

Example 4.2

An inverting summing amplifier has $R_{\text{in}-1} = 1.0$ kΩ, $R_{\text{in}-2} = 2.0$ kΩ, and $R_{\text{fb}} = 1.5$ kΩ. $V_1 = 2.0$ volts, and $V_2 = -2.0$V. What is the output voltage?

$$V_{\text{out}} = \frac{R_{\text{fb}}}{\dfrac{V_1}{R_{\text{in}-1}} + \dfrac{V_2}{R_{\text{in}-2}} + \dfrac{V_3}{R_{\text{in}-3}} + \ldots + \dfrac{V_n}{R_{\text{in}-n}}}$$

$$= \frac{1.5\text{ k}\Omega}{\dfrac{2.0\text{ V}}{1.0\text{ k}\Omega} + \dfrac{-2.0\text{ V}}{1.5\text{ k}\Omega}} = -2.25\text{ V}$$

NONINVERTING AMPLIFIER

A noninverting amplifier is shown in Fig. 4.4.

The output voltage and the resistors R_1 and R_2 make a voltage divider with its output at the inverting input $V_{\text{in}-}$. The voltage at that point is

$$V_{\text{in}-} = \left(\frac{R_1}{R_1 + R_2}\right) V_{\text{out}} \qquad 4.7$$

Figure 4.4 Noninverting Amplifier

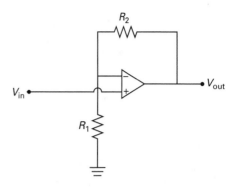

Since $V_{\text{in}+}$ is nearly equal to $V_{\text{in}-}$ (i.e., $V_{\text{in}+} \approx V_{\text{in}-}$),

$$V_{\text{out}} = \left(\frac{R_1 + R_2}{R_1}\right) V_{\text{in}+} \qquad 4.8$$

$$V_{\text{out}} = \left(\frac{R_1 + R_2}{R_1}\right) V_{\text{in}} \qquad 4.9$$

The gain is

$$A_V = \frac{V_{\text{out}}}{V_{\text{in}}} = \frac{R_1 + R_2}{R_1}$$

$$= 1 + \frac{R_2}{R_1} \qquad 4.10$$

The gain can never be less than one.

Sometimes a resistor is added between V_{in} and the non-inverting op amp input for temperature stabilization. Students upon first seeing this resistor may be startled, but very little current flows through this resistor so it does not affect gain calculations.

Example 4.3

The following circuit has resistances of $R_1 = 1.2$ kΩ and $R_2 = 6.2$ kΩ. What is the circuit gain?

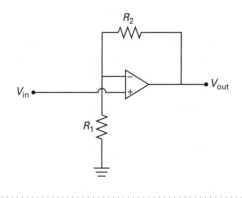

Solution

$$A_V = \frac{V_{\text{out}}}{V_{\text{in}}} = \frac{R_1 + R_2}{R_1}$$

$$= \frac{1.2 \text{ k}\Omega + 6.2 \text{ k}\Omega}{1.2 \text{ k}\Omega}$$

$$= 6.17$$

The circuit gain is independent of op-amp gain.

Voltage Follower

A *voltage follower* is made by modifying a noninverting amplifier like the one in Fig. 4.4. The resistance R_2 is changed to 0 Ω, and the resistance R_1 is changed to an open circuit, as shown in Fig. 4.5.

Figure 4.5 Voltage Follower

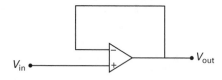

Sometimes a resistor is added between V_{in} and the noninverting op amp input for temperature stabilization. Students upon first seeing this resistor may be startled, but very little current flows through this resistor so it does not affect gain calculations.

The gain equation for a voltage follower is a modified version of the gain equation for the noninverting amplifier. The gain for a voltage follower is

$$A_V = \frac{V_{\text{out}}}{V_{\text{in}}}$$

$$= \frac{R_1 + R_2}{R_1}$$

$$= \frac{R_1 + 0 \ \Omega}{R_1}$$

$$= 1.0 \hspace{3cm} 4.11$$

The voltage follower is not useful as a high-gain voltage amplifier. It is used to match Thevenin high output-resistance signals to low-resistance loads. Op amp inputs use a negligible amount of current, so a voltage follower is an excellent method of measuring the voltage on a Thevenin high output-resistance circuit. The output voltage follows the signal input voltage. Indeed, this is why this particular op amp circuit is called a voltage follower. Op amps have a Thevenin low resistance output suitable for driving low-resistance loads. Field effect transistor (FET) op amp input currents are so low as to be extremely difficult to measure.

Crystal microphones and crystal piezoelectric transducers have high output resistances. Voltage followers or noninverting amplifiers are ideal for crystal microphones and transducers.

DIODES

Diodes are semiconductor devices that allow electrical current to travel in only one direction. Current is blocked from traveling in the reverse direction. A mechanical analogy is the check valve. The graph of current versus voltage in a silicon diode is shown in Fig. 4.6.

Figure 4.6 Current versus Voltage in a Diode

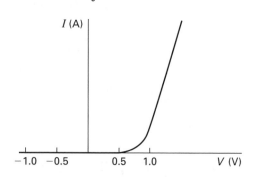

As the illustration shows, the silicon diode will begin to conduct above a threshold voltage of 0.7 V. The slope above 0.7 V is caused by internal resistance. Semiconductor diodes have a small reverse leakage current in the order of picoamperes.

Rectifiers, which convert AC to DC, use diodes. A diode is shown in the *half-wave rectifier* in Fig. 4.7. In this rectifier, the positive half cycle of an AC sinusoidal waveform passes through the diode in the arrow's direction and on to the resistor. The negative half cycle is blocked. That is why this circuit is called a half-wave rectifier.

Figure 4.7 Half-Wave Rectifier

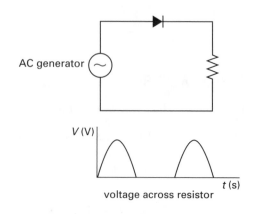

Four diodes can be used to make a *full-wave bridge rectifier* circuit, as shown in Fig. 4.8.

Figure 4.8 *Full-Wave Bridge Rectifier*

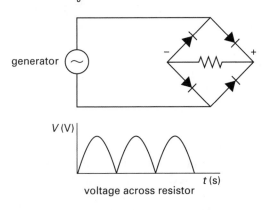

Figure 4.9 shows the direction of current flow in a full-wave bridge rectifier when the generator output is positive.

Figure 4.9 *Current Flow During a Positive Half Cycle*

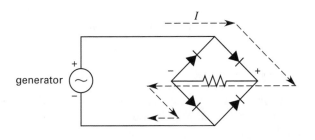

Figure 4.10 shows the current direction in a full-wave bridge rectifier during the generator's negative half cycle.

Figure 4.10 *Current Flow During a Negative Half Cycle*

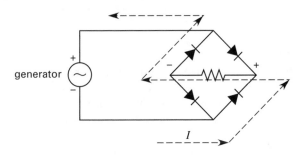

Another full-wave rectifier circuit is shown in Fig. 4.11. During the generator's positive cycle, the top diode conducts. During the negative half cycle, the lower diode conducts.

Figure 4.11 *Full-Wave Center-Tapped Rectifier*

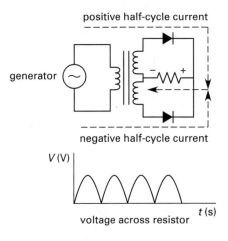

Inductive Protection Diodes

The voltage created across an inductor is

$$e = -L\frac{\Delta I}{\Delta t} \qquad 4.12$$

Inductive loads such as relays and solenoids are driven by mechanical switches and transistor switches. When these switches go to their open (non-conducting) states, the current quickly (i.e., the change in time is almost 0 s) ceases to flow, and an inductive voltage transient is generated. The voltage can easily damage transistor switches and eventually cause mechanical switch contacts to pit and corrode. A diode can be connected to catch the inductive transient and prevent damage to the switch.

In Fig. 4.12, the direction of inductor current is from top to bottom. The inductive transient is negative at the inductor top and positive at the inductor bottom. This polarity will drive the diode into conduction and safely dissipate the inductive transient, preventing damage to the switch.

Figure 4.12 *Diode Protection Circuit*

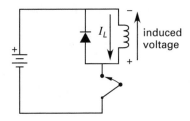

FILTERING

The pulsating DC voltages in the circuits described in the preceding Diodes section are suitable for motors, relays, and similar electro-mechanical devices, but not electronic circuits. For instance, a pulsating DC waveform powering an audio amplifier would produce an extremely annoying hum.

A capacitor is usually placed across the rectifier output, as shown in Fig. 4.13, to smooth the DC waveform. The capacitor stores charge when current flows through the diode, then releases the charge to the load resistor when the diode is not conducting. The smoothed waveform is shown in Fig. 4.14.

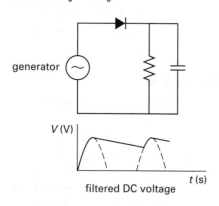

Figure 4.13 Filter for Pulsating DC Voltage

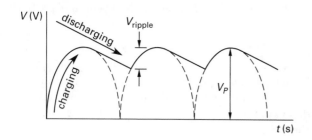

Figure 4.14 Ripple

The ripple voltage is the difference between the peak voltage and the minimum voltage. The percentage ripple is the ratio of the ripple voltage to the peak voltage.

$$\text{percentage ripple} = \frac{V_{\text{ripple}}}{V_P} \times 100\% \qquad 4.13$$

Full-wave rectification gives a smoother ripple and a lower percentage ripple as illustrated in Fig. 4.15.

The full-wave rectifier is charging the capacitor during each half cycle. The half-wave rectifier only charges the capacitor during the positive half cycle, and no charging occurs during the negative half cycle. The full-wave

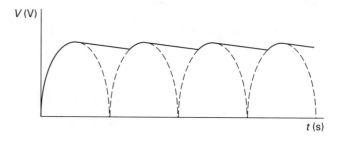

Figure 4.15 Ripple in a Full-Wave Capacitor Filter Rectifier

rectifier charges the capacitor and drives the load resistor during both half cycles. Thus the full-wave rectifier has less ripple than the half-wave rectifier. The amount of ripple is given by the following equations, in which $V_{\text{PP ripple}}$ is the peak-to-peak ripple voltage, f is the ripple frequency in hertz, R_L is the load resistance in ohms, C is the filter capacitance in farads, $V_{P\ \text{rectifier}}$ is the rectifier peak voltage output, and V_{DC} is the filter's DC output voltage.

$$V_{\text{PP ripple}} \approx \frac{1}{fR_LC}V_{P\ \text{rectifier}} \qquad 4.14$$

$$V_{\text{DC}} = \left(1 - \frac{1}{2fR_LC}\right)V_{P\ \text{rectifier}} \qquad 4.15$$

Example 4.3

The peak voltage from a half-wave rectifier with a capacitor filter is 18.6 V. The peak ripple voltage is 4.2 V. What is the percentage ripple?

Solution

$$\begin{aligned}
\text{percentage ripple} &= \left(\frac{V_{\text{ripple}}}{V_P}\right) \times 100\% \\
&= \left(\frac{4.2\ \text{V}}{18.6\ \text{V}}\right) \times 100\% \\
&= 22.6\%
\end{aligned}$$

Example 4.4

A full-wave bridge rectifier circuit has a capacitor filter and a peak voltage of 28 V, a capacitance of 47 μF, and a load resistance of 3.0 kΩ. The line frequency is 60 Hz. What is the output voltage and ripple voltage?

Solution

The potential trap in this problem for an unwary engineer is the frequency. The line frequency is 60 Hz, and the output of a half-wave rectifier is 60 Hz, but the output of a full-wave rectifier is 120 Hz. Both half cycles are used, so the filter frequency is twice the line frequency.

$$V_{PP \text{ ripple}} \approx \frac{1}{fR_LC}V_{P \text{ rectifier}}$$

$$\approx \left(\frac{1}{(120 \text{ Hz})\left(3 \times 10^3 \text{ } \Omega\right) \times \left(47 \times 10^{-6} \text{ F}\right)}\right)(28 \text{ V})$$

$$\approx 1.7 \text{ V}$$

$$V_{DC} \approx \left(1 - \frac{1}{2fR_LC}\right)V_{P \text{ rectifier}}$$

$$\approx \left(1 - \frac{1}{(2)(120 \text{ Hz})\left(3 \times 10^3 \text{ } \Omega\right) \times \left(47 \times 10^{-6} \text{ F}\right)}\right)(28 \text{ V})$$

$$\approx 27.2 \text{ V}$$

VOLTAGE REGULATORS

Often capacitor filters alone do not sufficiently smooth the DC voltage waveform. Inductors with capacitors are an additional option for filters, and voltage regulators provide excellent voltage regulation and filtering. Semiconductor integrated-circuit voltage regulators are available, inexpensive, and easy to use. A three-terminal regulator can simply be connected to the circuit after the capacitor filter. An engineer who installs the regulator only needs to make certain it is used correctly. Figure 4.16 shows how it fits into a circuit.

Figure 4.16 Integrated-Circuit Voltage Regulator

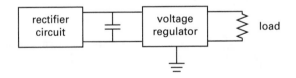

The part numbers for voltage regulators begin with 78 for positive voltages, and begin with 79 for negative voltages. For instance, the part number for a positive voltage regulator with a 12 V output would be 7812. The part number for a −5 V output regulator would be 7905. There are usually other manufacturer numbers and letters added to these four digits, but the basic part number is the same for all manufacturers.

PRACTICE PROBLEMS

1. An inverting op amp circuit has the values $R_{fb} = 22 \text{ k}\Omega$ and $R_{in} = 5.1 \text{ k}\Omega$. What is most nearly the circuit gain?

(A) −4.3
(B) −0.23
(C) 4.3
(D) 5.3

2. A noninverting op amp circuit has values of $R_1 = 5.1 \text{ k}\Omega$ and $R_2 = 22 \text{ k}\Omega$. What is most nearly the circuit gain?

(A) −4.3
(B) 1.0
(C) 4.3
(D) 5.3

3. The peak output voltage from a capacitor filtered rectifier is 24.6 V, and the peak ripple is 2.2 V. What is the percentage ripple?

(A) 0.089%
(B) 0.11%
(C) 8.9%
(D) 11%

4. A full-wave rectifier's peak 110 V output is filtered by a 120 μF capacitor. The line frequency is 50 Hz. The load is 2.4 kΩ. What is the approximate output ripple?

(A) 4.0 mV
(B) 2.7 V
(C) 3.8 V
(D) 7.6 V

SOLUTIONS

1. The circuit gain for this inverting op amp circuit is

$$A_V = -\frac{R_{fb}}{R_{in}} = -\frac{22 \text{ k}\Omega}{5.1 \text{ k}\Omega}$$
$$= -4.314 \quad (-4.3)$$

Note: Option (B) incorrectly inverts R_{fb} and R_{in} in the gain equation. Option (C) fails to recognize that this circuit has an inverting configuration. Option (D) also fails to recognize this is an inverting circuit, and uses an equation that finds the gain in a noninverting configuration.

Answer is A.

2. The circuit gain is

$$A_V = \frac{V_{\text{out}}}{V_{\text{in}}} = \frac{R_1 + R_2}{R_1}$$

$$= 1 + \frac{R_2}{R_1}$$

$$= 1 + \frac{22 \text{ k}\Omega}{5.1 \text{ k}\Omega}$$

$$= 5.314 \quad (5.3)$$

Note: Option (A) makes the mistake of using the equation for inverting circuits. Option (B) is the gain of a voltage follower, but this circuit is not a voltage follower. Option (C) applies the equation for gain incorrectly, leaving out the addition of 1.

Answer is D.

3.
$$\text{percentage ripple} = \left(\frac{V_{\text{ripple}}}{V_P}\right) \times 100\%$$

$$= \left(\frac{2.2 \text{ V}}{24.6 \text{ V}}\right) \times 100\%$$

$$= 8.94\% \quad (8.9\%)$$

Note: Option (A) does not multiply by 100%. Options (B) and (D) are the inverses.

Answer is C.

4. The output ripple is

$$V_{\text{ripple}} \approx \frac{1}{fR_LC} V_{P \text{ rectifier}}$$

$$\approx \left(\frac{1}{\substack{(100 \text{ Hz})\left(2.4 \times 10^3 \ \Omega\right) \\ \times \left(120 \times 10^{-6} \text{ F}\right)}}\right)(110 \text{ V})$$

$$\approx 3.82 \text{ V} \quad (3.8 \text{ V})$$

Note: Option (A) did not use the resistor and capacitor exponents properly. Option (B) incorrectly uses the rms value of the rectifier output voltage. Option (D) uses the line frequency of 50 Hz instead of the bridge output frequency of 100 Hz.

Answer is C.

5 Computer Hardware

Nomenclature

f	frequency	Hz
t	time	s
T	period	ns
V	voltage	V

DATA UNITS

The bit is the smallest unit of data. It is either true or false. Digital logic can only recognize voltage levels, so a true signal translates into 5 V, and a false signal translates into 0 V. A true takes the value of 1, and a false takes a value of 0. These ones and zeroes make binary number systems possible. Chapter 6 will show how bits can be used to count in the binary number system.

Eight bits of data constitute a byte. Four bits of data are a nibble, which can be thought of as a small byte. A word is 16 bits, 32 bits, or 64 bits, depending on the particular system.

In computer jargon, the prefix "kilo" (e.g., kilobit and kilobyte) does not mean exactly 1000. Rather, kilo means $(2)^{10} = 1024$. Similarly, "mega" (e.g., megabyte) means $(2)^{20} = 1\,048\,576$, and "giga" means $(2)^{30} = 1\,073\,741\,824$.

HOW DOES A COMPUTER WORK?

A digital computer performs instructions in a fetch-and-execute cycle. Program instructions (software) are stored in the computer's memory. The computer's *central processing unit* (CPU) fetches, or reads, one instruction from the computer's memory. The CPU decodes the instruction to determine what to do, and executes that instruction. The process is repeated until the program is finished.

The number of instruction types is surprisingly small. The instructions can be categorized as follows.

- read and write tasks involving internal memory
- read and write tasks involving peripherals (keyboard, monitor, modem, removable disks, etc.)
- arithmetic operations (add, subtract, multiply, divide)
- logical operations (this is covered in Ch. 6)
- comparison of two pieces of data, and decision making based on that comparison

Advanced computers can have less than 100 instructions in their repertoire. An *arithmetic logic unit* (ALU) needs only about 16 or 32 logical operations (shift the data, complement the data, perform OR, AND, EXOR instructions) to perform its tasks. Yet these few instructions are the building blocks of sophisticated software programs.

Comparing two pieces of data and making a decision based on that comparison is a powerful capability. The ALU compares two pieces of data to find which one is greater or whether they are equal. The program will alter its program sequence according to the program instruction.

The reason computers are part of our daily life is this ability to fetch and execute instructions and make millions of decisions per second. All this is done without fatigue, mistakes, coffee breaks, or labor strikes. This is what computers do. All the beautiful graphics and number crunching that the typical user sees is based on this simple premise. Complex software and programs are based on simple principles. Chapter 6 goes into greater detail on software.

COMPUTER ARCHITECTURE

The CPU is the part of the computer where the actual computing takes place. Data enter the CPU via various input devices, including scanners, keyboards, and sensors. Processed data leave via output devices such as monitor, printers, and plotters. Data and programs are stored in the computer memory. Data are exchanged with external systems via the internet using communications devices such as modems.

Figure 5.1 shows a block diagram of a typical computer. Figure 5.2 shows a more detailed block diagram of a computer.

Figure 5.1 Basic Computer Block Diagram

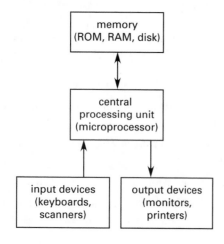

Buses

There are two configurations of buses: serial and parallel. A *serial bus* has one wire (line or trace). In this configuration, only one data piece can be sent at a time. Many communication systems are serial. Communication systems must connect towns and countries—long distances compared to buses that are part of a computer's internal workings—and it would be too costly to have parallel lines.

Figure 5.2 Computer Block Diagram

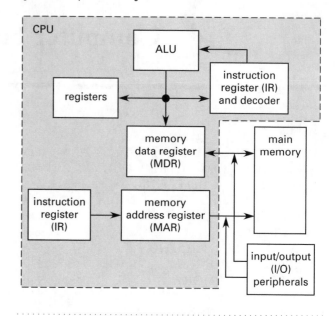

Parallel buses have several parallel lines instead of just one. Common sizes are 4, 8, 16, 32, or 64 lines (wires or traces). Nibbles, bytes, or words can be sent at once. This increased capacity makes a parallel bus much faster than a serial bus, and makes the parallel bus an ideal configuration for use inside a computer. The distances are short and the increased speed is easily obtained.

A computer has three buses (groups of wires or connections) that connect its various components. The three connections are control, address, and data buses. The *control bus* carries the timing, memory, and peripheral control signals among the CPU, memory, and I/O peripherals.

The *address bus* carries address signals to and from memory and peripherals. It is a parallel bus. Typical sizes (widths) are 8 bits, 16 bits, 32 bits, or 62 bits. A memory with 256 memory locations (addresses) will require an 8-bit address bus (i.e., $(2)^8 = 256$). A 16-bit address bus can catalog 65,536 addresses (i.e., $(2)^{16}$). Each address is a binary number. The signals are sent simultaneously on the parallel lines.

The *data bus* is a parallel bus carrying data between the CPU, memory, and peripherals. It works in conjunction with the address bus. The address bus sends an address to a memory device or peripheral device, and that device returns data back via the data bus. It is typically 4 bits, 8 bits, 16 bits, 32 bits, or 64 bits wide.

Instructions

A typical instruction cycle goes through the following steps if the instruction involves a memory or a peripheral.

1. The *memory address register* (MAR) places the instruction address onto the address bus.

2. The main memory reads the address, determines the address is meant for itself, and determines the address location within itself.

3. The main memory accesses the data at that address.

4. The main memory places the data onto the data bus.

5. The *memory data register* (MDR), a memory within the CPU, reads and stores the data.

6. The *instruction register* (IR) decodes the instruction within the data and activates the appropriate hardware to perform the instruction.

7. The process begins again.

If the instruction requires a comparison and decision, the process is modified. A typical comparison instruction cycle follows.

1. The MAR places the instruction address onto the address bus.

2. The main memory reads the address, determines the address is meant for itself, and determines the address location within itself.

3. The main memory accesses the two pieces of data to be compared.

4. The main memory places the data onto the data bus.

5. The MDR reads and stores that data.

6. The IR decodes the comparison instruction, and the ALU executes the instruction and compares the two data.

7. Based on whether one of the data is greater or lesser or whether they are equal, the next instruction will either be the next one in the program, or the MAR will be given a different address for the next instruction. This different address is within the compare instruction.

8. The process begins again.

Clock Cycles

The computer clock controls the timing. Data transfers and instructions are carried out in synchronism with the clock. Clock and timing signals are sent on the control bus.

A *microinstruction* is an instruction carried out in one clock cycle. Some instructions require more than one clock cycle. Main memory accesses are particularly slow and can take a large number of clock cycles.

Clock speeds of several billion cycles per second are now common. The ALU is specially designed to utilize these fast clock speeds. Other parts of the computer cannot be made to run as fast, but they will stay in synchronism with the main clock. The computer is thus a *synchronous machine*.

The clock signal is a square wave. Typical clock periods are on the order of 1 ns (10^{-9} s, or one nanosecond) or less. A clock's period, T, can be described with the following equation, in which f represents clock frequency.

$$T = \frac{1}{f} \qquad\qquad 5.1$$

BRANCHING INSTRUCTION

The comparison-and-decision instructions are also called *branching instructions*. Depending on the comparison result, the program flowchart may divert along branches to another task or program at a different address. Figure 5.3 shows a typical branching program flow for a data comparison in which the computer determines a number is either greater than zero (positive), less than zero (negative), or equal to zero.

The first comparison outcome causes the program to jump to task A if the number is positive. If the number is not positive, the program continues to the second comparison. If the number is less than zero, the program jumps to task B. If the number is not negative, it must be zero, and task C is executed. Tasks A and B reside in a different part of the main program than task C.

Figure 5.4 shows a typical application of branching instructions. An engineer pays a bill with a check. The check reaches the banking system computer and enters the main program. The computer compares the check amount to the engineer's checking account balance. If the account is overdrawn, a notice is sent to the engineer via a separate subprogram. If the balance is zero, the engineer receives a warning via another subprogram and a payment is made. If the balance is sufficient to cover the check, the main program makes payment and no warning is sent to the engineer.

Figure 5.3 Branching Program Flowchart

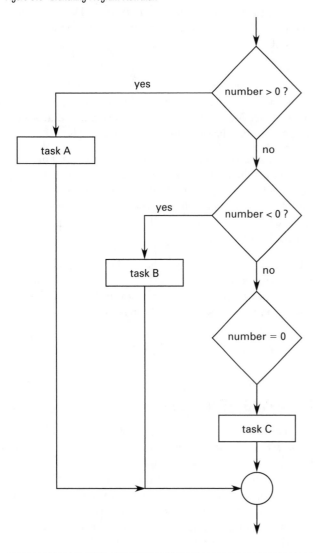

Figure 5.4 Example of Branching Instructions

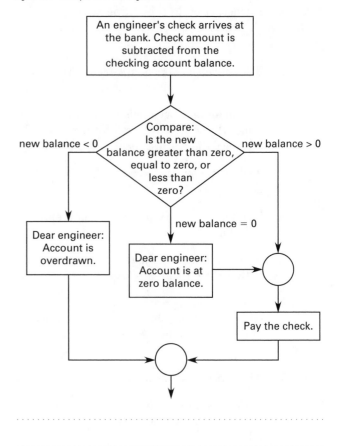

Figure 5.5 Program Storage and Access

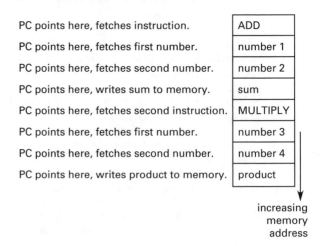

ACCESS TO PROGRAM STORAGE

Program storage refers to the method that the program is stored in main memory. The program counter is part of the MAR. Figure 5.5 shows how the program counter (PC) points to a memory address containing an ADD instruction. The instruction is read out of memory and sent to the MDR. The IR decodes the instruction as being an ADD instruction and determines that the next two addresses are needed for the two numbers. The two numbers are accessed from memory, sent to the ALU to be added, and their sum is written to memory.

The PC is incremented and points to the next instruction. The IR decodes it as a MULTIPLY instruction, which needs the numbers stored in the next two consecutive addresses. The PC points to the new numbers. They are brought into the MDR and then sent to the

ALU to be multiplied. The product of the numbers is sent to the MDR and finally written to memory.

COMPUTER MEMORY

Computer memory falls into two categories: volatile and nonvolatile. *Volatile memory* holds data as long as power is supplied to the memory. *Nonvolatile memory* keeps its data after the power is removed.

Volatile Memory

A volatile memory is a semiconductor memory. There are two types of volatile memory: dynamic random access memory (DRAM) and static random access memory (SRAM). *DRAM* stores data as charge on small capacitors. DRAM has a large capacity. It is cheaper than SRAM, but it is difficult to interface within the computer and difficult to manufacture. When power is removed, the charge on the capacitor leaks away. DRAM usually needs special circuits to maintain the capacitor charge. This process is called *refreshing*.

SRAM stores each bit on a small six-transistor circuit. SRAM is more expensive than DRAM, but it is also faster. SRAM is more expensive because each memory cell must be larger than a DRAM cell. Each type of volatile memory has its place within a computer system.

Nonvolatile Memory

Magnetic Memory

Magnetic memory stores data on magnetic media, usually disks coated with a ferrite material. Figure 5.6 shows a write coil for a magnetic disk (or platter). The current in the coil creates a magnetic field, which magnetizes a small part of the disk and forms a magnetic domain. The disk is spinning at high speed (typically 7200 rpm). As the disk rotates, the magnetic field changes direction as needed to write either a north pole or south pole into the ferrite.

Figure 5.6 Magnetic Disk Write Coil

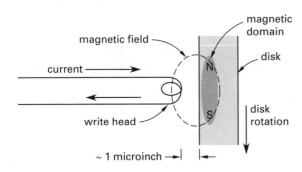

Figure 5.7 shows a voltage waveform read from a magnetic disk. The disk is rotating at a high speed (7200 rpm is typical). The magnetic domains and their fields move quickly past the read-head coil. The moving magnetic domains induce a voltage into the coil. This miniscule voltage is amplified and made available to the memory bus.

Figure 5.8 illustrates a rotating magnetic disk. The disk rotates past the arm with the read-write head. The head

Figure 5.7 Magnetic Disk Read Coil and Voltage Waveform

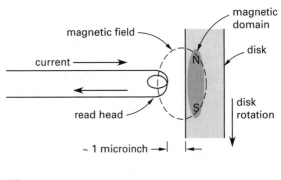

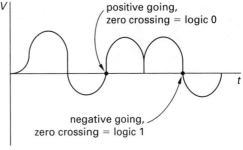

Figure 5.8 Magnetic Disk

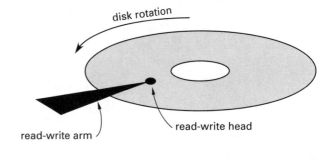

flies about one microinch above the disk, supported by a thin film of air. Driven by a step motor, the arm moves back and forth over the disk. The read-write head can reach any point on the moving disk.

Figure 5.9 shows the top view of a magnetic disk. The disk is divided into radial *sectors*, and the sectors are divided into concentric *tracks*. All the programmer sees are the memory addresses. The memory electronics take care of decoding, writing, and reading data.

Figure 5.10 shows a side view of a section of a multidisk storage system. Two magnetic disks are on the same rotating spindle. Each disk has two read-write arms moving in and out. Magnetic data is stored on both sides of each disk.

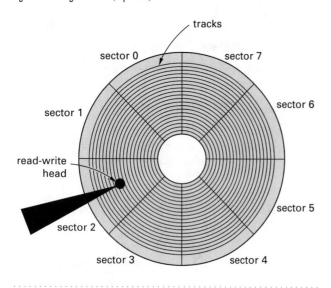

Figure 5.9 Magnetic Disk (Top View)

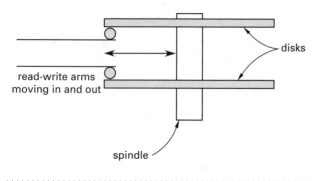

Figure 5.10 Multi-Disk Magnetic Storage (Side View)

Magnetic Tape Memory

Magnetic tape is used for archival storage. It is inexpensive, but access times are very slow. Nine tracks run along the tape, eight tracks for data, and a ninth track for parity error detection. There are nine stationary read-write heads. The write and read processes are essentially the same as for disk storage, except the heads do not move. Only the tape moves. The magnetic domains in the moving tape induce a voltage in the stationary read-head coil.

MEMORY USES

Besides simple data storage, memory applications in computers and other peripherals include registers, buffers, counters, caches, and first-in, first-out memories (FIFOs).

A *register* is temporary storage. It is usually 4 bits, 8 bits, 16 bits, or 32 bits wide. Data is stored until sent to

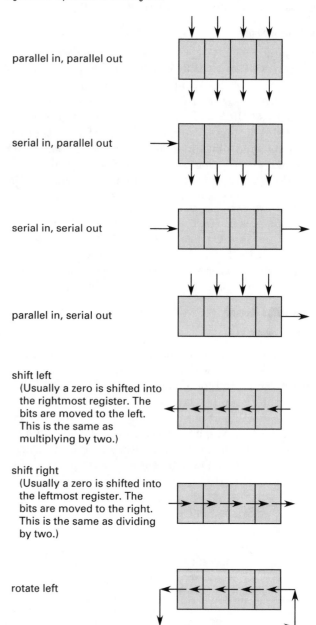

Figure 5.11 Operations of 4-Bit Registers

another unit, shifted, or rotated. Figure 5.11 shows the various operations of a 4-bit register. Each rectangular cell represents a 1-bit storage device.

A *buffer* provides temporary storage. It can be a register. Retrievals from magnetic storage, communication devices, and peripherals are relatively slow and usually not timed to the computer clock. Buffers hold the data until the transfer is complete and the computer or the peripheral can use the data.

A *counter* is a register that contains a digital number. The number can be incremented or decremented.

Cache memory is memory on the CPU. Main memory access is extremely slow compared to the CPU speed. The CPU will often have to wait a long time for data or instructions. Cache memory reduces the delay by reading data and instructions when the CPU is not performing a memory access. Often CPUs have two cache memories: one for instructions and the other for data. When the CPU needs the data or instructions, it gets them from the cache much faster than via a normal main memory access. Cache memories are usually SRAM because its speed is higher than DRAM.

FIFO (First In-First Out) memory provides an interface between systems of different speeds, such as between a computer and a communication system. A computer can perform millions of operations per second and transfer large blocks of data. Many communication systems can only operate at a fraction of that speed and must send one data unit at a time.

If the computer sends a block of data to a FIFO for transfer to a communications system, the FIFO passes on the data at a much slower rate to the communications system. The FIFO also aids the data transfer in the opposite direction. The communications system can send data one bit at a time into a FIFO. When the FIFO is nearly full, it signals the computer. The computer can then quickly read the accumulated data.

Another application of FIFO is reading from disk memory. A relatively long time is needed for the disk to rotate under the arm and for the read-write head to get into position. This means the access time for the disk is unpredictable. When finally read, the data comes quickly and is sent to a FIFO. It would waste a lot of time if the computer had to wait for the disk data to arrive. Instead the computer waits until the FIFO is nearly full and then reads the data quickly.

COMPUTER COMMUNICATIONS

Protocols

Communication systems have rules that enable them to function. These rules are called *protocols*. The rules include standards for speed, voltage levels, data formats, error handling, and even mechanical dimensions.

Recommended Standard RS-232

Recommended standard RS-232 is a serial communications protocol used for short distances between computer and peripherals. It uses a minimum of three lines: one for data and two for control.

Under this protocol, the device attempting to send data asserts a request-to-send ($\overline{\text{RTS}}$) signal. The bar over $\overline{\text{RTS}}$ signifies that the assertion changes from a high

voltage level (between 3 V and 15 V) to a negative voltage level (between −3 V and −15 V). The receiving device is able to receive the data and returns a clear-to-send ($\overline{\text{CTS}}$) signal. This signal is similarly asserted by changing from a high voltage to a negative voltage level. The sending device receives the $\overline{\text{CTS}}$ signal and begins sending data. When all the data have been transmitted, the sending device unasserts the $\overline{\text{RTS}}$ signal, and the transmission ceases. Figure 5.12 shows how RS-232 functions.

Figure 5.12 Recommended Standard RS-232

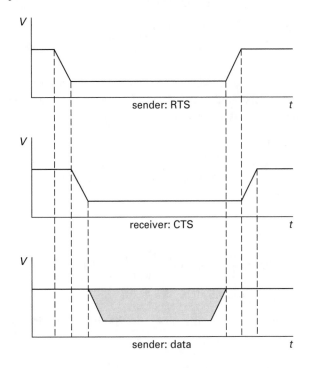

Simplex, Half-Duplex, and Full-Duplex Transmission

There are three modes of transmission: simplex, half duplex, and full duplex. *Simplex transmission* is a one-way communication transmission between a sending node and a receiving node. Examples of this transmission mode are radio and television broadcast stations.

In *half-duplex transmission*, one node transmits at a time while other nodes listen. Examples are public services such as police, fire departments, and taxi services. Amateur (ham) radio and citizen band radio are also half-duplex transmissions.

Full-duplex transmission allows nodes to transmit and listen simultaneously. The telephone system, with the proper name Public Switched Telephone Network (PSTN), is an example of full-duplex transmission. Most modern communication systems are full duplex.

Example 5.1

A 923 kilobyte (923 K) file is to be transmitted on a modem operating at 32.2 kilobits/s (32.2 Kbps). How much time will it take to transmit that file?

Solution

The file size is

$$(923 \text{ kilobytes}) \left(\frac{1024 \text{ bits}}{1 \text{ kilobyte}} \right) \left(\frac{8 \text{ bits}}{1 \text{ byte}} \right)$$

$$= 7\,561\,216 \text{ bits}$$

The modem transmits bits at 32 200 bps. The time required for the transmission is

$$\frac{7\,561\,216 \text{ bits}}{32\,200 \frac{\text{bits}}{\text{s}}} = 235 \text{ s}$$

Note: If these answers are rounded, they must be rounded up. It is impossible for a transmission to take less time than the precise answer.

Example 5.2

A 2.2 megabyte file is to be transmitted on a modem operating at 32.2 Kbps. How much time will it take transmit that file?

Solution

The file size is

$$(2.2 \text{ megabytes}) \left(\frac{1\,048\,576 \text{ bits}}{1 \text{ megabyte}} \right) \left(\frac{8 \text{ bits}}{1 \text{ byte}} \right)$$

$$= 18\,454\,938 \text{ bits}$$

The modem transmits bits at 32 200 bps. The time for the transmission is

$$\frac{18\,454\,938 \text{ bits}}{32\,200 \frac{\text{bits}}{\text{s}}} = 574 \text{ s}$$

Multiplexing

Multiplexing techniques allow several users to make use of the same transmission medium. Typical transmission mediums are twisted-pair copper wires, copper wires, optical fibers, satellites, and radio systems.

There are two types of multiplexing: *time division multiplexing* (TDM) and *frequency division multiplexing* (FDM). In TDM, each user shares (divides) time on a transmission medium. Each user is assigned a time slot to transmit data. The data is stored in a FIFO storage register. When the user's time slot arrives, the FIFO data is quickly read and sent onto the transmission medium. The first use of TDM was in the Wall Street financial area of New York. The existing telephone infrastructure had reached capacity, and the solution was TDM.

FDM users share frequency spectrum. Cable television and DSL fast internet access are examples.

Error Control

Invariably errors occur in computers and communications systems. There are two basic types of error control: *error detection* and *error correction*. Both require extra bits, called overhead.

Parity is the easiest error detection system to implement. An extra bit is added to a data byte or word. This extra bit tells whether the number of ones in the byte or word is odd or even. Odd parity requires the total number of ones to be odd. The parity bit will be set to make the total number of ones, including the parity bit, odd. Even parity requires the total number of bits to be even.

Parity can only detect one bit in error. This is satisfactory for checking computer memory errors, because they are usually random and not common. Communication system errors, however, occur in large blocks, so parity does not work well. Instead, the mathematical algorithm cyclic redundancy check (CRC) is used. When an error is detected, the system asks the sending unit to repeat the message.

Data read from a magnetic disk can be rife with flaws, and it is difficult to get a repeat of the data read. So error correction is used instead. Error correction is also better than error detection for applications such as deep space probes, which require a long time for a signal to travel from the spacecraft to the earth. The request for a repeat can take hours to perform. Error correction codes are used. All are based on the Reed-Solomon algorithm, a complete discussion of which goes beyond the scope of this book.

Example 5.3

Given the data byte shown, what would be added as the odd parity bit?

0 0 1 1 1 0 1 0

Solution

There are four 1s in the data byte, an even number. Thus, the parity bit must be a 1 for the total number of bits to be odd. The parity bit would be a 0 if the protocol were even parity.

Example 5.4

Given the data byte below, what what would be added as the even parity bit?

1 1 1 0 1 0 1 1

Solution

The total number of 1s is six, an even number. Thus the even parity bit must be a 0 to keep the total number of bits even. Thus, the complete byte and parity bit becomes

1 1 1 0 1 0 1 1 0

This is a total of 9 bits.

Synchronous and Asynchronous Communication

The Greek word *chronos* means time. *Synchronous* describes an activity that is based on a master clock. In data transfer, synchronous communication means the receiver knows precisely when the data will arrive. Synchronous data are sent in large data blocks. Each block has control bits that help the receiver synchronize with the incoming data and the clock on which the data is based. Error control bits arrive at the end of the data block.

With *asynchronous communications*, the receiver does not know when the data are coming, and the data are not well synchronized to a clock. RS-232 is an asynchronous protocol. It has a start bit, a stop bit, and perhaps a parity bit. The start bit prepares the receiver for the data.

LOCAL AREA NETWORKS

Local area networks (LANs) connect computers and peripherals together in a local area. The local area can be an office, building, campus or a home. Ethernet is a popular LAN system. Usually coaxial cable or twisted-pair copper wire connects the devices. However, wireless systems are becoming popular because installation is easier and faster, and straightforward to change. Wireless systems can be hacked much more easily than wired systems.

LANs allow *distributed processing*, meaning many machines can run one program simultaneously. Each machine does a portion of the program under the direction of the main program. This is a divide-and-conquer approach to complex programs with lots of data that would greatly burden a single machine.

INTERNET

The *internet* is a loosely organized system of computer networks spanning the globe. The internet includes the World Wide Web. The internet was designed by the United States military but has been given over to the civilian world. Through the internet, the home or office computer has access to information around the world. The internet also allows distributed processing.

PRACTICE PROBLEMS

1. What is a bit?

   (A) basic unit to encode a single text or numeric character
   (B) binary digit representing true or false
   (C) basic unit of computer data
   (D) both options (B) and (C)

2. How many bits constitute a byte?

   (A) 1
   (B) 2
   (C) 4
   (D) 8

3. How many bits constitute a nibble?

   (A) 1
   (B) 2
   (C) 4
   (D) 8

4. Which best describes the role of the control bus?

   (A) to send clock signals to the peripherals
   (B) to send clock and memory control signals to the memory
   (C) to send clock and READ, WRITE, and WAIT control signals to many computer areas
   (D) to receive control signals from peripherals

5. Program instructions are stored in the

   (A) CPU
   (B) main memory
   (C) ALU
   (D) volatile memory

6. What is the best definition of a microinstruction?

   (A) instruction executed in one clock cycle
   (B) small portion of an instruction
   (C) smallest time to perform an instruction
   (D) instruction executed by the ALU

7. What is the best description of a serial bus?

   (A) used by many communication systems
   (B) transmits multiple bits simultaneously
   (C) transmits one bit at a time
   (D) both options (A) and (C)

8. What is the function of the program counter?

(A) to count the number of programs the computer has executed

(B) to count the number of programs left to be executed

(C) to point to the addresses of instructions that have been executed

(D) to point to the address of the next instruction to be executed

9. Which descriptions best fit dynamic random access memory?

(A) volatile, inexpensive to make, relatively slow, needs to be refreshed

(B) nonvolatile, inexpensive to make, relatively fast

(C) volatile, expensive to make, fast

(D) nonvolatile, expensive to make, fast

10. What does a register do?

(A) permanently stores data

(B) temporarily stores data

(C) counts the number of program instructions

(D) provides memory that is always volatile

11. What is cache memory?

(A) nonvolatile memory

(B) dynamic memory

(C) memory that improves computer speed

(D) interface between CPU and communication peripherals

12. What is FIFO?

(A) nonvolatile memory

(B) dynamic memory

(C) memory that improves computer speed

(D) interface between CPU and communication peripherals

13. What is the best description of full duplex?

(A) simultaneous transmission between sender and receiver

(B) the public telephone system

(C) sender and receiver take turns transmitting

(D) both options (A) and (B)

14. What is multiplexing?

(A) performing many CPU tasks simultaneously

(B) accessing multiple memory devices

(C) multiple users sharing a transmission medium

(D) multiple users employing the same computer

SOLUTIONS

1. The definitions in options (B) and (C) are both correct. A bit is a basic unit of computer data, a binary digit that represents true and false.

Answer is D.

2. Eight bits make up a byte

Answer is D.

3. Four bits make up a nibble.

Answer is C.

4. The control bus sends clock, READ, WRITE, and WAIT control signals.

Answer is C.

5. Program instructions are stored in the main memory.

Answer is B.

6. A microinstruction is an instruction that a computer executes in one clock cycle.

Answer is A.

7. A serial bus is described by both options (A) and (C). A serial bus transmits one bit at a time and is used by many communication systems.

Answer is D.

8. The function of a program counter is to point the computer to the address of the next instruction to be executed.

Answer is D.

9. DRAM is inexpensive to make, but it is relatively slow, volatile, and needs to be refreshed.

Answer is A.

10. A register temporarily stores data.

Answer is B.

11. Cache memory is memory on the CPU that reduces delays associated with main memory access by reading data and instructions when the CPU is not performing a memory access.

Answer is C.

12. FIFO memory provides an interface between systems of different speeds, such as between a computer and a communication system.

Answer is D.

13. Options (A) and (B) are both correct. Full duplex transmission is simultaneous communication between sender and receiver. This is the way the public telephone system works.

Answer is D.

14. Multiplexing is a transmission mode in which many users share a transmission medium.

Answer is C.

6 Computer Software

Nomenclature

n number of comparisons   –
N number of entries   –

SOFTWARE

Software comprises the programs and operating system of a computer. The computer itself and accessories such as monitor, keyboard, and mouse are the *hardware* discussed in Ch. 5.

The *operating system* is the interface between the user and the computer. The operating system schedules and runs the programs, and controls the hardware. Modern operating systems such as Windows, Linux, and OS-X hide the details of the machine from the user, making the computer easier to use. All the user does is point and click with the mouse, and the software and hardware respond by doing amazing things. The user does not see the complicated process between the mouse click and the machine's action.

The *programs* tell the machine and operating system what to do. If the program is stored on disk or tape, it is called software. Examples are word-processing and spreadsheet programs. If the program is stored in the computer's read-only memory (ROM), it is called *firmware* (halfway between hardware and software). The ROM can be easily changed as updates become available. Firmware examples are boot-up routines and some operating system functions.

PROGRAM DESIGN

A program is a series of instructions sent to the machine that enable it to do a task or function. Examples of instructions are WRITE, READ, and GOTO. These types of instructions, called *source code*, can be read by humans, but the machine needs to have them translated into machine language, machine code, or *binary code*. The machine can only read voltage levels, so the code is only a series of different combinations of the numerals one (a high voltage signal) and zero (a low voltage signal).

Programmers write the source code, and the source code is translated into machine code or an executable program. Special programs do this translation, which means programmers have programs to help them program.

FLOWCHARTS

A *flowchart* is a diagram showing how a program proceeds. A flowchart aids in understanding the program, what the program does, and the various steps and paths the program execution can take. Flowcharts can be used for other processes, such as mail delivery, manufacturing steps, check processing, and food product distribution. Often a programming error becomes obvious by careful examination of the flowchart. Figure 6.1 shows a selection of flowchart symbols. Figure 6.2 shows how these symbols might be used in a simple program.

Figure 6.1 Flowchart Symbols

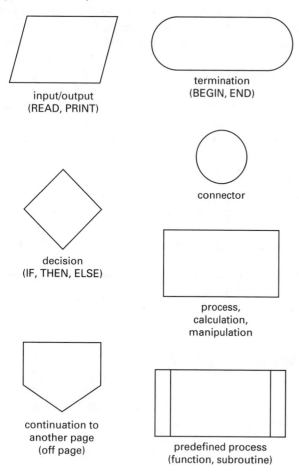

Find roots of $aX_2 + bX + c = 0$ using the following quadratic equation.

$$X = \frac{-b \pm \sqrt{b^2 - 4ac}}{2a}$$

LANGUAGE LEVELS

Programming languages can be categorized by their level. What determines a level is who or what reads the program (i.e., human programmer or machine) and how close the program is to machine language.

Low-Level Language

A computer can only understand high voltages or low voltages. These voltages are interpreted as binary code:

a 1 bit or 0 bit, respectively. These ones and zeroes are represented by a high voltage and a low voltage respectively. A program that consists of these ones and zeros is called a *low-level language* or *machine language*. All programs must be translated into a low-level language for the machine to execute the program. Such a program is called *an executable program* or *executable file*.

Figure 6.2 Program Flowchart for Quadratic Formula Solutions

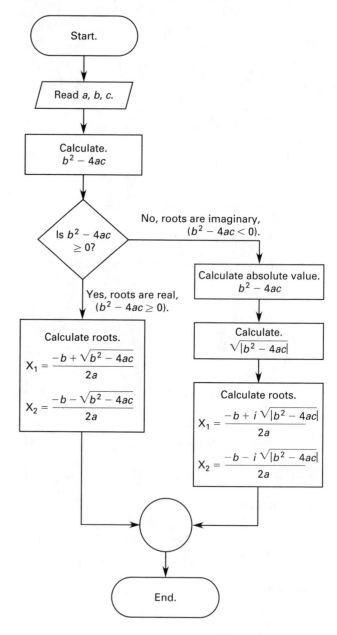

Each instruction is a set of ones and zeros called *intrinsic machine code*. There are two parts to the instruction, the *op-code* and the *operand*. The op-code tells the computer what operation is to be performed, and the operand is the information with which the computer works. For instance, the op-code for an instruction to add numbers might be 11011001, and the operands (the numbers to be added) could be the binary numbers 11100010 and 10001001. Often the operands are stored in memory, so the instruction contains the op-code and the memory addresses for the numbers to be added.

Assembly Language

It is possible for a human to write machine language code, but this is tedious, error prone, and prohibitively inefficient. Higher level languages are much more efficient and allow greater programmer creativity and ingenuity. *Assembly language* programs use *mnemonic statements* that represent machine level instructions. Mnemonic statements are English-like commands that describe the instruction. For instance, the following assembly language command tells the computer to subtract the quantity in memory location A from the quantity in memory location B and to store the result in memory location B.

SUB A, B

The mnemonic SUB stands for subtract. A and B are the operands (quantities to be operated upon). Note that the memory locations are given as the variables A and B, rather than in machine language. The computer's instruction register decodes the ones and zeroes and determines that a subtraction is to occur, and then tells the memory address register (MAR) where the operands A and B are located in memory.

These types of statements are much easier for humans to comprehend, but the computer cannot understand assembly code. The assembly program must be translated into machine language code. What better way to do this boring and error-prone task than to let the computer itself make the translation? The assembly language program is given to an *assembler*, which assembles (translates) the program into machine language code. Assemblers are more than mere translators, however. They also assign addresses to the operands, handle program branches, and catch syntax errors.

A microprocessor is basically a central processing unit (CPU). The terms are similar and often used interchangeably. However, larger computers have CPUs but don't have a microprocessor. Microprocessors are used in personal computers and smaller workstations.

Each microprocessor has its own assembly language. For instance, a program written for the Intel microprocessors will not work with Motorola microprocessors. Assembly language programs are therefore not *portable*. However, Intel's assemblers are quite similar within the company's family of microprocessors, and differ only in improvements and additions.

High-Level Language

What seems to be a trivial task can require many statements of an assembly language program. A simple request to print the file RESUME might consist of more than 50 assembly language statements. A multiply or divide instruction can also have many statements. It would be inefficient if programmers had to write this type of code into each of their programs.

High-level languages use shorthand commands such as the following.

PRINT RESUME

$Y = W*X/Z$

The translation from these high-level commands to machine language is performed either by an *interpreter* program or a *compiler* program. A compiler translates the entire program, checks syntax, and outputs an *executable program* (a stand-alone, ready-to-run, machine code program).

An interpreter checks syntax, translates, and executes one line of code at a time. It does not produce a stand-alone program. Interpreters are good for debugging a program, but are much slower than compilers. Also, the translation and syntax checks must be done each time the program runs.

High-level languages include C, C++, FORTRAN, BASIC, Visual Basic, and LISP. High-level languages from all software companies will ideally run on all brands of machine: PCs, Macs, and the different mini-computers and mainframe computers. This ideal of portability is generally not the case in reality, but it usually takes only minimal rewriting to make code work on different machines. Graphics and display issues usually require the most rewriting.

Obviously high-level languages are easier for humans to work with than assembly languages. Code can be written much faster. However, the compiled executable program usually does not run as fast as a carefully written assembly language program. Some executable programs take days or weeks to run. For instance, programs exist that can accurately predict the weather for a relatively long time into the future. Unfortunately, these programs are so huge and run so slowly that a weather prediction for the next three days takes over a week to execute. In such a situation, writing certain program portions in assembly language is worthwhile.

Memory management and saving memory space are also easier to do with assembly programs. If memory space is critical, a well crafted assembly language program may save critical memory.

Structured Language

Early programming languages such as FORTRAN and COBOL used many GOTO statements. These statements caused the program sequence to jump to and from various program segments. The unfortunate result was code that was difficult to follow and understand. The programmer himself often had difficulty following the program flow after several weeks away from the program. Bitter programmers called these programs spaghetti code, meaning following the program flow was like following spaghetti noodles through a plate of pasta.

Structured programming languages, such as C, C++, and Pascal, divided the program into parts that are variously called functions, subroutines, or procedures. The ideal main program would have two portions: the program parts and a series of calls to these parts. Many of the procedures are supplied by a library of subprograms. These languages are self documenting, which make the programmer's life much easier. C and C++ have GOTO statements, but these are considered bad form and are only used as a last resort.

Variables used only in the subprogram are called *local variables*. Global variables can be used by the main program and all subprograms.

C and C++ permit a function to call itself. Examples are computing the nth power of 2, Fibonacci numbers, and factorials. These are called *recursive functions*.

The most common and tricky questions are loops counts. Calculating the number of loops is a prime area for mistakes. The best procedure for these problems is

- Determine what the program does.
- Find the counting loop.
- Find the initial count.
- Find where the count is incremented or decremented.
- Do a hand count to accurately determine the final count or number of times the loop is invoked. A hand count is laborious, but reduces the chances of an error in these very tricky problems.

Example 6.1

Determine the number of times the loop is executed.

```
A = 12
DO IF A < 3 LEAVE PROGRAM
A=A−2
END DO
```

Solution

A	LEAVE PROGRAM
12	no
10	no
8	no
6	no
4	no
2	yes

The program passes through the loop five times and leaves the loop on the sixth pass.

Example 6.2

How many loops does this WHILE program perform?

```
X=10
WHILE (X > 0)
    X = X−1
END WHILE
```

Solution

X	END
10	no
9	no
8	no
7	no
6	no
5	no
4	no
3	no
2	no
1	no
0	yes

The program performs the loop ten times.

CODING

Binary Coding

A digital computer uses voltage levels to distinguish numbers. Only two voltage levels are used, typically 0 V for a "logic 0" (i.e., false) and a positive 5 V for a "logic 1" (i.e., true). (The true and false are also used in Boolean algebra.) Because the computer can only recognize two states or levels, a binary (base 2) number system is used.

The binary bit is the smallest piece of data. It can only have two values, 0 (false) or 1 (true). A byte is an 8-bit grouping. Half of a byte (4 bits) is a nibble. A word is either a 16-bit, a 32-bit, or a 64-bit grouping, depending on the exact application and machine.

The decimal (base 10) system provides a useful starting point for understanding how the binary system works. In the decimal system, each digit to the left multiplies the numeral in that spot by an increasing power of 10. Two examples follow.

$$(247)_{10} = (2)(10)^2 + (4)(10)^1 + (7)(10)^0$$
$$= (2)(100) + (4)(10) + (7)(1)$$
$$= 200 + 40 + 7$$
$$= 247$$
$$(803)_{10} = (8)(10)^2 + (0)(10)^1 + (3)(10)^0$$
$$= (8)(100) + (0)(10) + (3)(1)$$
$$= 800 + 0 + 3$$
$$= 803$$

The computer uses a binary (base 2) system in a similar manner. Two examples follow.

$$(13)_{10} = (1)(2)^3 + (1)(2)^2 + (0)(2)^1 + (1)(2)^0$$
$$= (1)(8) + (1)(4) + (0)(2) + (1)(1)$$
$$= 8 + 4 + 0 + 1$$
$$= (1101)_2$$

The most significant bit (MSB) is the leftmost bit, and is a 1 with the value 8. The least significant bit (LSB) is the rightmost bit, and has a value of 1.

$$(54)_{10} = (1)(2)^5 + (1)(2)^4 + (0)(2)^3$$
$$+ (1)(2)^2 + (1)(2)^1 + (0)(2)^0$$
$$= (1)(32) + (1)(16) + (0)(8) + (1)(4)$$
$$+ (1)(2) + (0)(1)$$
$$= 32 + 16 + 0 + 4 + 2 + 0$$
$$= (110110)_2$$

The MSB is the leftmost bit with a value of 32. The LSB is the rightmost bit with a value of 0.

A more elegant method of converting decimal to binary is the divide-by-two method. Under this method, the decimal number is divided by two. The whole number result and the remainder (either one or zero) are recorded. Then the whole number result is again divided by two, and the new whole number result and the new remainder are set aside. The process repeats itself until the whole number result is finally either one or two and that result is divided by two. When the process is complete, the binary number appears: in the remainder with the bit order from bottom to top. In the following demonstration, $(13)_{10}$ is converted to its binary equivalent using this method.

$$13/2 = 6, \text{ remainder} = 1$$
$$6/2 = 3, \text{ remainder} = 0$$
$$3/2 = 1, \text{ remainder} = 1$$
$$1/2 = 0, \text{ remainder} = 1$$

The binary bit order in the remainder is from bottom to top so

$$(13)_{10} = (1101)_2$$

The binary equivalent of $(13)_{10}$ is $(1101)_2$. When this method is used the most significant bit is always a 1.

Example 6.3

Convert $(247)_{10}$ to its binary equivalent using the divide-by-two method.

Solution

$$247/2 = 123, \text{ remainder} = 1$$
$$123/2 = 61, \text{ remainder} = 1$$
$$61/2 = 30, \text{ remainder} = 1$$
$$30/2 = 15, \text{ remainder} = 0$$
$$15/2 = 7, \text{ remainder} = 1$$
$$7/2 = 3, \text{ remainder} = 1$$
$$3/2 = 1, \text{ remainder} = 1$$
$$1/2 = 0, \text{ remainder} = 1$$

The binary number bit order is from bottom to top in the remainder, so

$$(11110111)_2 = (247)_{10}$$

The binary equivalent of $(247)_{10}$ is $(11110111)_2$.

Octal Coding

The binary system is at the computer/machine level. But using this system is painfully slow and error prone for humans. Other number base systems are used to aid human readability: the base 8 or octal system, and the base 16 or hexadecimal system.

Two examples of the base 8 system follow.

$$(13)_{10} = (1101)_2$$
$$= (1)(8)^1 + (5)(8)^0$$
$$= (15)_8$$
$$(54)_{10} = (110110)_2$$
$$= (6)(8)^1 + (6)(8)^0$$
$$= (66)_8$$

Hexadecimal Coding

As data buses and address buses became larger, octal representation also became difficult for humans to read. Hexadecimal is easier for people to comprehend, but it requires practice. The hexadecimal system uses the first six letters of the alphabet to represent the numbers 10 through 15. The following table shows examples of numbers represented by the hexadecimal system and three commonly used number systems.

Table 6.1 Four Number Systems

decimal	binary	octal	hexadecimal
0	0 0 0 0 0 0 0 0	0 0 0	0 0 0
1	0 0 0 0 0 0 0 1	0 0 1	0 0 1
2	0 0 0 0 0 0 1 0	0 0 2	0 0 2
3	0 0 0 0 0 0 1 1	0 0 3	0 0 3
4	0 0 0 0 0 1 0 0	0 0 4	0 0 4
5	0 0 0 0 0 1 0 1	0 0 5	0 0 5
6	0 0 0 0 0 1 1 0	0 0 6	0 0 6
7	0 0 0 0 0 1 1 1	0 0 7	0 0 7
8	0 0 0 0 1 0 0 0	0 1 0	0 0 8
9	0 0 0 0 1 0 0 1	0 1 1	0 0 9
10	0 0 0 0 1 0 1 0	0 1 2	0 0 A
11	0 0 0 0 1 0 1 1	0 1 3	0 0 B
12	0 0 0 0 1 1 0 0	0 1 4	0 0 C
13	0 0 0 0 1 1 0 1	0 1 5	0 0 D
14	0 0 0 0 1 1 1 0	0 1 6	0 0 E
15	0 0 0 0 1 1 1 1	0 1 7	0 0 F
16	0 0 0 1 0 0 0 0	0 2 0	0 1 0
17	0 0 0 1 0 0 0 1	0 2 1	0 1 1
254	0 1 1 1 1 1 1 0	3 7 6	0 F E
255	0 1 1 1 1 1 1 1	3 7 7	0 F F
256	1 0 0 0 0 0 0 0	4 0 0	1 0 0
257	1 0 0 0 0 0 0 1	4 0 1	1 0 1
258	1 0 0 0 0 0 1 0	4 0 2	1 0 2
511	1 1 1 1 1 1 1 1	7 7 7	1 F F

Character Coding

In the binary system, alphanumeric data (the letters A through Z) and numeric data (0 through 9) are represented by a series of ones and zeros. There are two commonly used codes to denote alphanumeric data: the American Standard Code for Information Interchange (ASCII) and the Extended Binary Coded Decimal Interchange Code (EBCDIC).

ASCII (pronounced "ask-kee") is a 7-bit code shown in Table 6.2. Seven bits allow $(2)^7 = 128$ combinations. This code includes control characters and alphanumeric symbols. Some control characters are holdovers from Teletype uses. ASCII is the standard for microcomputers.

There is an 8-bit version of ASCII, with Greek letters and mathematical symbols. Eight bits allow $(2)^8 = 256$ combinations.

EBCDIC (pronounced "eb-sih-dik") was developed by IBM. It is an 8-bit code used in large computers. IBM chose to use ASCII for the PC, rather than EBCDIC.

LOGIC OPERATIONS

Digital *integrated circuits* (IC) can perform logic functions such as AND, OR, INVERT (NOT), and combinations such as NAND (AND and INVERT) and NOR (OR and INVERT). These ICs are called logic gates or simply gates. Most gates are a single stage inverting amplifier, and have both voltage and current amplification. Noninverting single stage amplifiers have only current amplification or voltage amplification, not both. Inverting amplifiers used in gates are easier to make and are faster than noninverting amplifiers. Op amps are much slower and more complicated than logic gates, and only rarely used for logic functions.

INVERT or NOT Gate

An INVERT or NOT function will change a true (logic 1) to false (logic 0) or a false (logic 0) to true (logic 1). That is, a high voltage (5 V = logic 1) on the inverter gate input will give a low voltage (0 V = logic 0) on the inverter gate output. A low voltage on the inverter gate input will give high voltage output. Table 6.3 is a truth table for the inverter.

The series 110010110101 has the inverse 001101001010. In another form, the NOT gate input and output is

$$110010110101$$
$$001101001010$$

AND Gate

An AND gate will give a true output only when all of its inputs are true. There is no theoretical limit to the number of inputs, but a practical limit is eight inputs. Table 6.4 is a two-input AND gate truth table.

The following shows two bit streams and their AND output.

$$110010110101$$
$$010101010101$$
$$010000010101$$

OR Gate

An OR gate will give a true output when one or more of inputs are true. Its truth table is described in Table 6.5.

The following shows two bit streams and their OR output.

$$110010010101$$
$$010101010111$$
$$110111010111$$

Table 6.2 ASCII

name	decimal	binary	hex	symbol	decimal	binary	hex	symbol	decimal	binary	hex	symbol	decimal	binary	hex
	control character				graphic symbol				graphic symbol				graphic symbol		
NUL	0	0000000	00	space	32	0100000	20	@	64	1000000	40	`	96	1100000	60
SOH	1	0000001	01	!	33	0100001	21	A	65	1000001	41	a	97	1100001	61
STX	2	0000010	02	"	34	0100010	22	B	66	1000010	42	b	98	1100010	62
ETX	3	0000011	03	#	35	0100011	23	C	67	1000011	43	c	99	1100011	63
EOT	4	0000100	04	$	36	0100100	24	D	68	1000100	44	d	100	1100100	64
ENQ	5	0000101	05	%	37	0100101	25	E	69	1000101	45	e	101	1100101	65
ACK	6	0000110	06	&	38	0100110	26	F	70	1000110	46	f	102	1100110	66
BEL	7	0000111	07	'	39	0100111	27	G	71	1000111	47	g	103	1100111	67
BS	8	0001000	08	(	40	0101000	28	H	72	1001000	48	h	104	1101000	68
HT	9	0001001	09	)	41	0101001	29	I	73	1001001	49	i	105	1101001	69
LF	10	0001010	0A	*	42	0101010	2A	J	74	1001010	4A	j	106	1101010	6A
VT	11	0001011	0B	+	43	0101011	2B	K	75	1001011	4B	k	107	1101011	6B
FF	12	0001100	0C	,	44	0101100	2C	L	76	1001100	4C	l	108	1101100	6C
CR	13	0001101	0D	-	45	0101101	2D	M	77	1001101	4D	m	109	1101101	6D
SO	14	0001110	0E	.	46	0101110	2E	N	78	1001110	4E	n	110	1101110	6E
SI	15	0001111	0F	/	47	0101111	2F	O	79	1001111	4F	o	111	1101111	6F
DLE	16	0010000	10	0	48	0110000	30	P	80	1010000	50	p	112	1110000	70
DC1	17	0010001	11	1	49	0110001	31	Q	81	1010001	51	q	113	1110001	71
DC2	18	0010010	12	2	50	0110010	32	R	82	1010010	52	r	114	1110010	72
DC3	19	0010011	13	3	51	0110011	33	S	83	1010011	53	s	115	1110011	73
DC4	20	0010100	14	4	52	0110100	34	T	84	1010100	54	t	116	1110100	74
NAK	21	0010101	15	5	53	0110101	35	U	85	1010101	55	u	117	1110101	75
SYN	22	0010110	16	6	54	0110110	36	V	86	1010110	56	v	118	1110110	76
ETB	23	0010111	17	7	55	0110111	37	W	87	1010111	57	w	119	1110111	77
CAN	24	0011000	18	8	56	0111000	38	X	88	1011000	58	x	120	1111000	78
EM	25	0011001	19	9	57	0111001	39	Y	89	1011001	59	y	121	1111001	79
SUB	26	0011010	1A	:	58	0111010	3A	Z	90	1011010	5A	z	122	1111010	7A
ESC	27	0011011	1B	;	59	0111011	3B	[	91	1011011	5B	{	123	1111011	7B
FS	28	0011100	1C	<	60	0111100	3C	\	92	1011100	5C	|	124	1111100	7C
GS	29	0011101	1D	=	61	0111101	3D	]	93	1011101	5D	}	125	1111101	7D
RS	30	0011110	1E	>	62	0111110	3E	^	94	1011110	5E	~	126	1111110	7E
US	31	0011111	1F	?	63	0111111	3F	_	95	1011111	5F	Del	127	1111111	7F

Table 6.3 NOT Truth Table

input	output
0	1
1	0

Table 6.4 AND Truth Table

input A	input B	AND output
0	0	0
1	0	0
0	1	0
1	1	1

Table 6.5 OR Truth Table

input A	input B	OR output
0	0	0
1	0	1
0	1	1
1	1	1

NAND Gate

A NAND gate is equivalent to an AND gate with an inverter (NOT gate) after the AND. The truth table is shown in Table 6.6.

Table 6.6 AND-NAND Truth Table

input A	input B	AND	NAND
0	0	0	1
1	0	0	1
0	1	0	1
1	1	1	0

From our AND example,

$$110010110101$$
$$010101010101$$
$$\overline{010000010101}$$

Based on that AND output, the NAND output is

$$101111101010$$

In practice, NAND gates are faster and cheaper than AND gates. The AND gate construction is actually a NAND gate followed by an inverter, which makes AND gates slower and more expensive.

NOR Gate

A NOR gate is equivalent to an OR gate with a NOT gate after the OR output. The truth table is described in Table 6.7.

Table 6.7 OR-NOR Truth Table

input A	input B	OR	NOR
0	0	0	1
1	0	1	0
0	1	1	0
1	1	1	0

From the OR example

$$110010010101$$
$$010101010111$$
$$\overline{110111010111}$$

Based on that OR output, the NOR output is

$$001000101000$$

In practice, NOR gates are cheaper and faster than OR gates. The OR gate construction is a NOR gate followed by an inverter, which makes the OR gate slower and more expensive.

It is possible to build any logic function using only NAND gates or only NOR gates. This usually is not the best way to design logic, but it can be done.

Exclusive OR and Exclusive NOR Gates

The truth table for an exclusive OR (EXOR) gate and an exclusive NOR (EXNOR) gate is described in Table 6.8.

Table 6.8 EXOR-EXNOR Truth Table

input A	input B	EXOR	EXNOR
0	0	0	1
1	0	1	0
0	1	1	0
1	1	0	1

Note that the EXOR output is true only when its inputs are unequal, and is false only when its inputs are equal. This makes an EXOR gate useful for digital comparators.

An EXOR output from two bit streams is

$$110010110101$$
$$010101010101$$
$$\overline{100111100000}$$

Based on that EXOR output, the EXNOR output is

$$011000011111$$

Another use for an EXOR gate is controlling whether a gate is an inverter or noninverter. A control signal bit stream and a signal bit stream has the following EXOR output.

control	11111111	00000000	11111111
signal	00110001	11011000	01110011
EXOR	11001110	11011000	10001100
	invert	noninvert	invert

Example 6.4

For the following logic expression, what is the output (value) if A = 1, B = 0, and C = 0?

(A.OR.B).AND.(NOT(C))

Solution

First find (A.OR.B).

(A.OR.B) = (1 OR 0) = 1

Then find (NOT(C)).

(NOT(C)) = (NOT(0)) = 1

Put the results together.

(A.OR.B).AND.(NOT(C))
  = (1.OR.0).AND.(NOT(0))
  = (1.AND.1)
  = 1

Example 6.5

For the following logic expression, what is the output (value) if A = 1, B = 0, and C = 1?

(A.AND.B).EXOR.(B.OR.C)

Solution

First find (A.AND.B).

(A.AND.B) = (1.AND.0) = 0

Then find (B.OR.C).

(B.OR.C) = (0.OR.1) = 1

Finally, put the expression together.

(A.AND.B).EXOR.(B.OR.C)
  = (1.AND.0).EXOR.(0.OR.1)
  = (0.EXOR.1)
  = 1

FILES, RECORDS, AND FILE TYPES

Fields are empty spaces on an electronic form that are designed to collect specific information, such as a person's name, address, and telephone number. A collection of this type of information about one person or one thing is called a *record*. A group of records is a *data file*. A file can be a program or data.

Sequential files are usually stored on magnetic tape, and must be read from the beginning. An *index* is typically used to aid in locating files on the tape.

Random file structures are stored on disks, allowing quick access to any sector and track. Records may be placed at different locations on the disks, not necessarily in sequence. Reading a complete file could require several jumps from one location to the next. There are means to reconstruct the file in sequential order.

A customer list is an example of a data file. A customer list is usually kept in the order entered. A new customer record is added to the end of the file, without regard to alphabetic order. The list of records could be sorted into alphabetic order each time a new customer is added, but this would be inefficient. An improved method of sorting the list might use an index table, which could be sorted according to last name. Each entry in the index table points to a particular customer field in the main customer list. Sorting the list can be done by fields other than last name, such as phone number, age, or zip code. Sorting is covered in more detail in the next section of this chapter.

Database

A *database* is a file that contains related information, such as a mailing list. The mailing list would have records such as name, street address, city, state, country, postal code, and telephone number. Databases are organized and accessed by different methods.

Flat File

A *flat file* has one key reference word for each record, and that key word is located within its record. The key word is entered into the database search algorithm, and the database is searched for a match.

Key and Data File

A *key and data file* system has two files. Key words are in one file, and the data is in the second file. Table 6.9 is an example of this type of file system.

Linked List

In a *linked list*, or *threaded list*, each record except the last one has a pointer to the next record in the list. The pointer is usually an address or record number. The sequence is established by the pointers and does not have to follow any other order, such as alphabetized names. Figure 6.3 is an example of a linked list.

Tree Structure

In a *tree structure* list, each record has pointers to other records that match specific criteria. Records are usually called *nodes* in a tree structure database. A binary tree structure is shown in Fig. 6.4. The number of comparisons, n, needed to find an entry can be calculated with the following equation, in which N is the number of entries.

$$n = 1 + \frac{\log N}{\log 2} \approx \log N \qquad 6.1$$

Table 6.9 Key and Data File System

key file

Adams	1
Lincoln	3
Washington	5
Roosevelt	4
Jackson	2

data file

1	Adams	John	2nd president
2	Jackson	Andrew	7th president
3	Lincoln	Abraham	16th president
4	Roosevelt	Theodore	26th president
5	Washington	George	1st president

Figure 6.3 Linked List

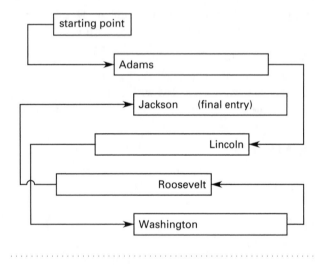

Figure 6.4 Binary Tree Structure

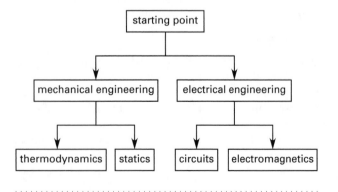

Hierarchical Structure

A *hierarchical structure* organizes each record by a main entry and includes subentries underneath the main entry. In Fig. 6.5, the main entry is a U.S. president's last name, and the subentries are a first name and the president's place in the line of presidents.

Relational Structure

A *relational structure* organizes its records in matrix form. Table 6.10 shows an example of relational structure. Information can be accessed by reference to a field name (e.g., record number, last name, first name, or presidential order) or to a field value (e.g., a search for presidential order = 16 will bring up the record for Lincoln).

Table 6.10 Relational Data Base

record no.	last name	first name	order
1	Adams	John	2
2	Jackson	Andrew	7
3	Lincoln	Abraham	16
4	Roosevelt	Theodore	26
5	Washington	George	1

SORTING

Sorting puts data into descending or ascending numerical order or into alphabetic order. Several sorting routines exist, ranging from a simple but slow sort to faster, more sophisticated *quicksort*, *insertion sort*, and *heap sort* methods.

Alphabetic sorts are done by comparing the ASCII codes.

A *bubble sort* compares each record entry to the following entry. If the bubble sort is placing records in ascending order and the second entry is larger, the two entries are swapped. The larger entry bubbles to the top of the list. If no swaps occur during a pass, the list is considered sorted. The number of comparisons in a bubble sort is described by

$$n = \frac{N^2}{2} \qquad 6.2$$

It is obvious that for large numbers of entries, the bubble sort is quite slow.

Figure 6.5 Hierarchical Structure

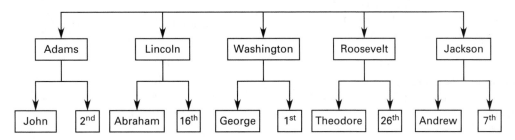

An insertion sort rewrites the entries in their proper order. An entry's correct placement is found, all entries below that place are pushed down, and the entry is inserted into the empty place. The average number of comparisons for this type of sort is

$$n = \frac{N^2}{4} \qquad 6.3$$

The worst-case number of comparisons for this sort is given by Eq. 6.3.

The bubble sort and the insertion sort become very slow when the number of comparisons is large and approaches the square of the number of entries ($n = N^2$).

When the number of entries is large, the quicksort algorithm is much faster but more complicated than the bubble sort or the insertion sort. The number of comparisons is approximately

$$n = \frac{N \log N}{\log 2} \approx \log N \qquad 6.4$$

If the original list is nearly sorted, the quicksort method loses its speed advantage.

The approximate number of comparisons for a heap sort is the same as for the quicksort.

Searching

Searching is the process of looking through a database for a certain record or records. A *linear search*, or *sequential search*, is useful for randomly organized records. It examines each record in turn. The average number of comparisons is half the number of entries ($N/2$), but it can reach the same number as the number of entries.

A *binary search* works best for records in ascending or descending order. The search finds the middle value of the list, narrows its search to the appropriate half (upper or lower), and repeats the process, which in effect cuts the number of relevant records in half at each round until the desired record is found. The comparisons needed are given by Eq. 6.4.

Hashing is a mathematical technique to speed up the searching process. A key word or value is converted to a record number by a *hashing algorithm*, or *hashing function*. A common method is division using only the remainder. A collision happens if there is another key with the same hashing record number, so chaining, double hashing, and linear probing are used to handle these problems.

ARTIFICIAL INTELLIGENCE

Artificial intelligence (AI) is the ability of a computer to learn concepts, reason, and output logical solutions. An example is a medical computer that could hold a symptom database, learn from new information, and give useful diagnoses.

In an actual case, a dam engineer was about to retire. His knowledge of the dam was going to be severely missed. AI experts interviewed him and entered his responses into their AI system and developed a workable system.

SPREADSHEETS

Electronic spreadsheets are an important tool for engineers and many other professionals. The FE spreadsheet problems usually give several numbers in cells and a formula in other cells. The examinee will be required to find a number in a cell using the given numbers and formula.

Microsoft Excel® is a commonly used spreadsheet program. Fill commands are used to save repetitive rewriting of data or formulas. The fill command can be a copy, linearly increasing or decreasing, date or growth, and can fill columns or rows.

An absolute reference is an equation's reference to a specific cell and does not change when the equation itself is copied or moved to another cell. An example is =$A$2, which references cell A2. Absolute references use dollar signs.

A relative reference is an equation's reference to a cell. The reference changes depending on the cell to which the equation is moved or copied. Relative references do not use dollar signs as do absolute references.

Example 6.6

In the spreadsheet, the formula =(A2^3)/3 is entered into cell B1 and copied into cells B2 to B7. What is the value in cell B4?

	A	B	C	D	E
1	2				
2	3				
3	4				
4	5				
5	6				
6	7				
7	8				
8					

Solution

The formula in cell B4 uses the value in cell A5, which is 6.

$$|B4| = \frac{(A5)^3}{3} = \frac{(6)^3}{3}$$
$$= 72$$

The formula changes when it is copied down column B, and the references are relative.

Example 6.7

In the following spreadsheet, the formula =$A$1*$A$3 is entered into cell B1 and copied into cells B2 to B7. What is the value in cell B1?

	A	B	C	D	E
1	2				
2	3				
3	4				
4	5				
5	6				
6	7				
7	8				
8					

Solution

The values in cell A1 (2) and in cell A3 (4) are multiplied to give an answer of 8. The cell references are absolute; that is, they reference specific cells and do not change with a copy or fill command. All cells B1 through B7 have the same answer.

PRACTICE PROBLEMS

1. Convert $(145)_{10}$ into a binary (base 2) number. Use the divide by two method.

    (A) $(10010001)_2$
    (B) $(10010001)_2$
    (C) $(111100)_2$
    (D) $(10001001)_2$

2. Convert $(145)_{10}$ into a hexadecimal (base 16) number. Use the results of Prob. 1.

    (A) $(FF)_{16}$
    (B) $(91)_{16}$
    (C) $(191)_{16}$
    (D) $(1616)_{16}$

3. Convert the binary number $(1010101)_2$ to a base 10 number.

    (A) $(65)_{10}$
    (B) $(69)_{10}$
    (C) $(85)_{10}$
    (D) $(101)_{10}$

4. Convert $(145)_{10}$ to an octal (base 8) number. Use the results of Prob. 1.

    (A) $(111)_8$
    (B) $(2F1)_8$
    (C) $(221)_8$
    (D) $(811)_8$

5. Convert the binary number $(101111010)_2$ to an octal number.

    (A) $(434)_8$
    (B) $(552)_8$
    (C) $(572)_8$
    (D) $(1350)_2$

6. What operation is typically represented by the programming flowchart symbol shown?

    (A) terminal
    (B) input/output
    (C) connector
    (D) decision

7. Given X = true, Y = true, and Z = false, what is the value of the following logical expression?

$$(X.AND.(NOT(Z)).OR.(Y)$$

(A) true
(B) false
(C) not true or false
(D) true or false

8. Convert the hexadecimal number DA2 into a binary number.

(A) $(001001011011)_2$
(B) $(110110100010)_2$
(C) $(101011010100)_2$
(D) $(111011101101)_2$

9. Convert the binary number $(111101010100)_2$ to an octal number.

(A) $(3459)_8$
(B) $(7512)_8$
(C) $(7524)_8$
(D) $(7528)_8$

10. What is a compiler?

(A) program to translate source code into an executable file
(B) program to convert assembly language into an executable file
(C) program to convert each line of source code one at a time and execute it
(D) program to handle input/output functions

11. What is an interpreter?

(A) program to translate source code into an executable file
(B) program to convert assembly language into an executable file
(C) program to convert each line of source code one at a time and execute it
(D) program to handle input/output functions

12. What is an assembler?

(A) program to extract source code from an executable file
(B) program to convert assembly language code into an executable file
(C) program to convert each line of source code one at a time and execute it
(D) program to handle input/output functions

13. How many bits constitute a nibble? A byte?

(A) 4 bits and 8 bits, respectively
(B) 8 bits and 4 bits, respectively
(C) 8 bits and 16 bits, respectively
(D) 16 bits and 32 bits, respectively

14. What does a recursive function do?

(A) evaluates residual functions
(B) calls itself
(C) controls memory accesses
(D) controls communication peripherals

15. Given the following pseudo-code, what is the final value of Y?

```
Y=0
X=3
DO REPEAT UNTIL X<0
X=X−1
Y=Y+1
END REPEAT
```

(A) −1
(B) 0
(C) 3
(D) 4

16. In the following spreadsheet, the formula =A1*A2 is entered into cell B1 and copied through column B. What is most nearly the value in cell B3?

	A	B
1	5	
2	4	
3	3	
4	2	
5	1	

(A) 3.0
(B) 6.0
(C) 140
(D) 400

17. In the following spreadsheet, the formula =A1^A2 is entered into cell C1 and copied through column C. What is most nearly the value in cell C4?

	A	B	C
1	5		
2	4		
3	3		
4	2		
5	1		

(A) 1.0
(B) 2.0
(C) 16
(D) 25

SOLUTIONS

1.
$$145/2 = 72, \text{ remainder } 1$$
$$72/2 = 36, \text{ remainder } 0$$
$$36/2 = 18, \text{ remainder } 0$$
$$8/2 = 4, \text{ remainder } 0$$
$$9/2 = 4, \text{ remainder } 1$$
$$4/2 = 2, \text{ remainder } 0$$
$$2/2 = 1, \text{ remainder } 0$$
$$1/2 = 0, \text{ remainder } 1$$

Reading the remainders from bottom to top, $(145)_{10} = (10010001)_2$.

Note: Option (D) is the top to bottom order.

Answer is B.

2. Use the binary equivalent for $(145)_{10}$, which as shown in the first solution is $(10010001)_2$, to find the hexadecimal equivalent.
$$(1001)_2 = (9)_{16}$$
$$(0001)_2 = (1)_{16}$$

Answer is B.

3.
$$(1)(2)^6 + (0)(2)^5 + (1)(2)^4 + (0)(2)^3$$
$$+ (1)(2)^2 + (0)(2)^1 + (1)(2)^0$$
$$= 64 + 0 + 16 + 0 + 4 + 0 + 1$$
$$= (85)_{10}$$

Answer is C.

4. The binary equivalent for $(145)_{10}$ is $(10010001)_2$, as found in Prob. 1. Splitting the binary number,
$$10 \quad 010 \quad 001 = (2)(8)^2 + (2)(8)^1 + (1)(8)^0$$
$$= (221)_8$$

Answer is C.

5. Split the binary number $(101111010)_2$ to obtain the octal number.
$$101 \quad 111 \quad 010 = (5)(8)^2 + (7)(8)^1 + (2)(8)^0$$
$$= (572)_8$$

Note: Check the octal number.
$$(5)(64) + (7)(8) + (2)(1) = (378)_{10}$$

Answer is C.

6. The parallelogram represents an input/output.

Answer is B.

7. For the given values of variables X, Y, and Z, the value of the logical expression is true.

$X = 1$

$Y = 1$

$Z = 0$

(X.AND.(NOT(Z)).OR.(Y)

X.AND.(NOT(Z))=(1.AND.1)=1

Answer is A.

8. Split the hexadecimal number DA2.
$$D = (1101)_2$$
$$A = (1010)_2$$
$$2 = (0010)_2$$

Therefore,
$$(DA2)_{16} = (110110100010)_2$$

Answer is B.

9. Split the binary number $(111101010100)_2$.
$$(111)_2 = (7)_8$$
$$(101)_2 = (5)_8$$
$$(010)_2 = (2)_8$$
$$(100)_2 = (4)_8$$

Therefore,
$$(111101010100)_2 = (7524)_8$$

Answer is C.

10. A compiler is a program to translate source code into an executable file.

Answer is A.

11. An interpreter is a program that converts each line of source code one at a time and executes it.

Answer is C.

12. An assembler is a program that converts assembly language into an executable file.

Answer is B.

13. Four bits constitute a nibble. Eight bits constitute a byte.

Answer is A.

14. A recursive function can call itself.

Answer is B.

15. The final value of Y is 4. The loop was executed four times.

X	Y	END
3	0	no
2	1	no
1	2	no
0	3	no
−1	4	yes

Answer is D.

16. The value in cell B3 is 6.0.

	A	B	C
1	5	A1*A2=5*4	
2	4	A2*A3=4*3	
3	3	A3*A4=3*2	
4	2	A4*A5=2*1	
5	1		

Answer is B.

17. The value in cell C4 is 2.0.

	A	B	C
1	5		A1^A2=$(5)^4$=625
2	4		A2^A3=$(4)^3$=64
3	3		A3^A4=$(3)^2$=9.0
4	2		A4^A5=$(2)^1$=2.0
5	1		

Answer is B.

7 Measurement and Instrumentation

Nomenclature

A	area	m^2
A	gain	–
c	speed of light	m/s
C	capacitance	F
d	displacement, distance,	
	spacing	m
E	error	%
I	current	A
l	length	m
Q	charge	C
V	voltage	V
T	temperature	°C, °F, K

Symbols

α	resistance temperature	
	coefficient	1/°C
ε	permittivity	F/m
ρ	resistivity	$\Omega \cdot$m

Subscripts

out	output
ref	reference
T	temperature

PURPOSE OF MEASUREMENT

Measurement is any process used to ascertain the state or condition of a physical or chemical quantity. With this knowledge, a machine or a human operator can determine that a quantity is what is desired and proceed with an engineering, scientific, or business mission. Or the operator can determine that the quantity is not what is desired and take corrective action.

OUTPUT TO MACHINE AND HUMAN

Physical quantities are rarely in a form that can be readily comprehended by a machine or its human operator.

Humans are unreliable at quantifying physical phenomena with their own senses. For instance, some people are quite comfortable at temperatures that mean misery to other people. People need devices to accurately and consistently tell them a number associated with a physical quantity such as temperature. These devices are called *instruments* or *transducers*.

Some transducers have outputs that can be easily read by humans. A bathroom scale has a needle that moves across a scale in response to weight against a spring. Many pressure gauges work similarly. Mercury in a thermometer moves up and down with the mercury's expansion and contraction with temperature. Barometric pressure is measured by the expansion or contraction of a sealed air chamber. The expansions or contractions are mechanically linked to an indicator needle.

Other transducers create electrical outputs that are related to the measured physical quantities. These outputs are easily read by a computer. There are many kinds of outputs for human observation, including voltmeter readings, frequency readings, and pressure gauge displays.

ANALOG-TO-DIGITAL AND DIGITAL-TO-ANALOG CONVERTERS

The input to an *analog-to-digital converter* (ADC) is usually a voltage, current, or frequency. The ADC output is a digital binary code easily read by a computer. The computer program will examine the ADC output and send a reading to an operator. Of course, this reading will not be in binary but is a display meaningful to the operator (e.g., numeric value for voltage, pressure, temperature, or pH).

The computer program may determine that some action is needed without an operator's intervention. The computer program's response may be to send a binary code to a *digital-to-analog converter* (DAC). The DAC output is usually a voltage or current. Sometimes the output will drive an analog meter for human observation, but often the output is fed back into the process to keep the process output at a desired value. An example is a thermostat keeping a temperature constant. Modern home thermostats are digitally controlled to vary temperature according to time of day and day of the week.

ADC and DAC Limitations

Accuracy and precision are critical to ADCs and DACs. They are limited by *resolution*. This is a function of how many bits (steps or states) are output from the ADC or how many bits drive the DAC.

$$\text{resolution} = \frac{\text{input range}}{2^n \text{ steps}} \qquad 7.1$$

For instance, if an 8-bit ADC input is 1.00 V maximum, the resolution is

$$\text{resolution} = \frac{\text{input range}}{2^n \text{ steps}}$$

$$= \frac{1.00 \text{ V}}{(2)^8 \text{ steps}}$$

$$= 0.0039 \text{ V/step} \quad (3.9 \text{ mV/step})$$

The conversion gain, A, for both types of converters is the output divided by the input. For ADC, the output is described as a binary code range and the input is in voltage, frequency, current, or similar units.

$$A_{\text{ADC}} = \frac{\text{output}}{\text{input}} \qquad 7.2$$

For DAC, the output is in units of voltage, frequency, current, or similar units, and the input is described as a binary code range.

$$A_{\text{DAC}} = \frac{\text{output}}{\text{input}} \qquad 7.3$$

Linearity is a measure of how much the gain changes over the full input range. In real converters each step is not identical, so the gain can vary from the ideal linear curve.

Monotonic or *monotonicity* means the output will increase with increasing input, or decrease with decreasing input. It is possible for an ADC or a DAC output to decrease instead of increase as expected. This problem generally only happens over a small range of inputs. This may not be a problem when the output is on a display.

However, control systems can easily become unstable with this type of error. This is a potentially catastrophic situation. The instrument in error might mistakenly signal the control system to increase its output, when the proper control system response is to decrease its output. The control system would be operating in reverse with possible disastrous results. Converters are usually specified as being monotonic or not.

Example 7.1

An 8-bit ADC has an input of 0.0–5.0 V. What is the output gain?

Solution

$$A_{\text{ADC}} = \frac{\text{output}}{\text{input}}$$

$$= \frac{(2)^8 \text{ steps}}{5.0 \text{ V}}$$

$$= 51.2 \text{ steps/V}$$

Example 7.2

A 12-bit DAC has an output of 0.0–1.0 mA. What is the conversion gain?

Solution

$$A_{\text{DAC}} = \frac{\text{output}}{\text{input}}$$

$$= \frac{1.0 \text{ mA}}{(2)^{12} \text{ steps}}$$

$$= 0.000\,244\,1 \text{ mA/step} \quad (0244.1 \text{ } \mu\text{A/step})$$

SIGNAL CONDITIONING

Transducer outputs are rarely in a form compatible with either computer inputs or meter displays for people. Special electronic circuits convert (condition) the transducer signal into a form a computer or a meter display can use.

Some signal-conditioning circuits convert the transducer signal to a 0–5 V output, compatible with many ADCs.

A commonly used signal-conditioning output is 4–20 mA. The 4 mA lower end of this range corresponds to a minimum sensor output, and the 20 mA upper end corresponds to a maximum sensor output. This current-based signal-conditioning system has built in failure detection. If the current drops to 0 mA, either the signal conditioner has failed or the wire connection has opened up or shorted. A current transmission system is more noise resistant than a voltage transmission system. Some ADCs operate directly from a 4–20 mA input, and some DACs directly output 4–20 mA.

ACCURACY, ERROR, AND PRECISION

Accuracy describes how close measurements are to the true value. Students using digital meters often take the readings as absolute truth; they think it has to be right because it is digital. Just because there are many digits on the readout does not mean the instrument is accurate to that number of digits. Digital instruments are not exempt from drift, age, temperature variations, and partial failures. Digital instruments still need calibration.

Error is how far the measurements are from the true value. The percentage error, E, of a measurement can be found with the following equation.

$$E = \left(\frac{\text{measured value} - \text{true value}}{\text{true value}} \right) \times 100\%$$

$$7.4$$

Precision is not the same thing as accuracy. Precision is the repeatability or agreement of measurements. An instrument could display the same value for 100 consecutive readings. This would be a very precise instrument. But those readings could be off by a fixed error, so the instrument is not necessarily accurate. Good precision does not equal good accuracy.

Example 7.3

A voltmeter reading is 12.000 V. The true value is 11.910 V. What is the percentage error?

Solution

$$\begin{aligned} E &= \left(\frac{\text{measured value} - \text{true value}}{\text{true value}} \right) \times 100\% \\ &= \left(\frac{12.000 \text{ V} - 11.910 \text{ V}}{11.910 \text{ V}} \right) \times 100\% \\ &= 0.75567\% \end{aligned}$$

THERMOCOUPLES

A *thermocouple* is a simple sensor that measures voltage between its wires to determine the temperature difference between two points. A thermocouple consists of two wires of dissimilar metals joined together at one end with an ammeter or voltmeter at the other end. The joined end (probe) can be twisted together, but soldering, brazing, or welding make a better mechanical connection. The probe is exposed to a temperature, T. The other end, the reference junction near the ammeter or voltmeter, is kept at a reference temperature, T_{ref}, usually maintained in ice water at 0°C. Electronics can simulate the ice bath.

Figure 7.1 Thermocouple

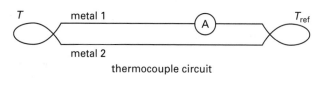

thermocouple circuit

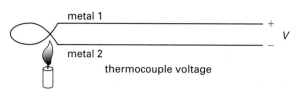

thermocouple voltage

A current flows if the temperature at the probe is different from the reference temperature because of the dissimilar metals in the thermocouple circuit (see Fig. 7.1). This is the *Seebeck effect*. The voltage generated is small, in the order of microvolts per degree of temperature difference.

The approximate voltage generated in the thermocouple is given by the following equation, in which V is the voltage and k_T is the first order thermocouple constant with units of microvolts per degree Celsius or Fahrenheit.

$$V = k_T(T - T_{\text{ref}})$$

$$7.5$$

The actual thermocouple constant is not a linear function as assumed in Eq. 7.5. A more accurate description uses a polynomial approximation, in which V is the thermocouple voltage, a is a polynomial coefficient unique to each thermocouple type, and n is the maximum polynomial order.

$$\begin{aligned} T &= a_0 + a_1 V + a_2 V^2 + a_3 V^3 + a_4 V^4 \\ &\quad + a_5 V^5 + \cdots + a_n x^n \end{aligned}$$

$$7.6$$

The polynomial coefficients are available from industrial tables or from careful calibration. Obviously polynomial calculations like this are best done with a computer. Thermocouple error is about 0.5–0.7%.

Table 7.1 lists the materials used in thermocouple wires, the American National Standards Institute (ANSI) designation for the different types of thermocouples, the first order thermocouple constant, and the useful temperature range. Constantan is a copper-nickel alloy, alumel is an aluminum-nickel alloy, and chromel is a chromium-nickel alloy.

Often the thermocouple measurement takes place at a great distance from the reference temperature and meter. The two thermocouple metals are typically only used for a short distance in the sensor, and these are connected to the meter with long copper wires. This is a cheaper solution than using expensive thermocouple metals such as platinum and rhodium across the whole distance. Characteristics of thermocouples and other transducers are summarized in Table 7.2.

Example 7.4

An ANSI type J thermocouple measures $500°C$ at the probe. The reference junction is in an ice bath at a temperature of $0°C$. What is the output voltage? Use the first order thermocouple constant.

Solution

$$
\begin{aligned}
V &= k_T \left(T - T_{\text{ref}}\right) \\
&= \left(51 \, \frac{\mu V}{°C}\right) \left(500°C - 0°C\right) \\
&= 25\,500 \, \mu V \quad (25.5 \text{ mV})
\end{aligned}
$$

RESISTANCE TEMPERATURE DETECTORS

Resistance temperature detectors (RTDs) make temperature measurements based on the change in metal's resistance versus temperature. Table 7.2 contains a summary of RTD characteristics.

Metal resistance variation with temperature was discussed in Ch. 1. The first order approximate equation for the relationship between resistance and temperature in RTDs follows, with R as the resistance, T_{ref} as the reference temperature, T as the measured temperature, and α as the coefficient of temperature (i.e., the change in resistance per temperature degree).

$$
\begin{aligned}
R_T &= R_{T_{\text{ref}}} \left(1 + \alpha \left(T - T_{\text{ref}}\right)\right) \\
&= R_{T_{\text{ref}}} \left(1 + \alpha \left(\Delta T\right)\right) \qquad \textit{7.7}
\end{aligned}
$$

The RTD is more linear than the thermocouple, but a quadratic curve fitting is used for improved accuracy over the preceding equation. In Eq. 7.8, α_1 is the linear coefficient of temperature and α_2 is the quadratic (second order) coefficient of temperature. The first order coefficients use units of $1/°C$ or $1/°F$, and the second order coefficients have units of $1/°C^2$ or $1/°F^2$.

$$
\begin{aligned}
R_T &= R_{T_{\text{ref}}} \left(1 + \alpha_1 \left(T - T_{\text{ref}}\right) + \alpha_2 \left(T - T_{\text{ref}}\right)^2\right) \\
&= R_{T_{\text{ref}}} \left(1 + \alpha_1 \left(\Delta T\right) + \alpha_2 \left(\Delta T\right)^2\right) \qquad \textit{7.8}
\end{aligned}
$$

This is best solved by a computer. A Wheatstone bridge, a type of strain gauge described later in this chapter, is often used for more accurate resistance measurement.

Table 7.1 Thermocouple Characteristics

materials for wire 1, wire 2	ANSI designation	thermocouple constant, k_T ($\mu V/°F$)	thermocouple constant, k_T ($\mu V/°C$)	useful temperature range (°F)	useful temperature range (°C)
copper, constantan	T	21	38	−300 to 700	−180 to 370
iron, constantan	J	28.3	51	32 to 1400	0 to 760
chromel, alumel	K	22.2	40	32 to 2300	0 to 1260
chromel, constantan	E	32.5	58.5	32 to 1600	0 to 870
platinum, 10% rhodium	S	5.7	10.3	32 to 2700	0 to 1480
platinum, 13% rhodium	R	6.4	11.5	32 to 2700	0 to 1480

Table 7.2 Transducers

type	principle	output	characteristics
linear variable differential transducer (LVDT)	Transformer core moves with displacement. The primary to secondary AC coupling depends on the core displacement. Output varies with the relative position of the core to the coils.	voltage phase and amplitude	• measures displacements of up to several inches (accuracy to micrometers)
variable capacitance transducer	One of the variables is changed, either distance between plates, capacitor area, or the relative dielectric constant. $$C = \frac{\varepsilon_r \varepsilon_0 A}{d}$$	frequency shift	• measures small displacement or chemical composition of dielectric
piezoelectric sensor	Pressure squeezes a piezoelectric crystal, creating an electric charge.	charge converted to voltage	• low output, much amplification needed • amplifier input must be very high impedance • only for fast-changing pressures because charge leaks away
photovoltaic sensor	Light falling on sensor generates an electrical voltage.	current, voltage	• limited spectral range • also called photoelectric cells
photoconductive sensor	Light falling on sensor changes resistance.	current, voltage	• limited spectral range • also called photoresistive sensor
photodiode sensor	Light falling on sensor releases electrons.	current	• used in optical communications systems
resistance temperature detector (RTD)	Temperature fluctuations change wire resistance; devices usually have positive temperature coefficient.	bridge output	• accurate temperature detection • can be used in harsh, high temperature environments
thermistor	Semiconductor temperature detector with negative temperature resistance coefficient.	voltage, current, bridge	• not very accurate; large unit to unit variation • used to stabilize temperature for electronic circuits
thermocouple	Voltage is generated across two wires of different metals as the thermocouple detects a temperature difference between its probe and a reference (Seebeck effect).	voltage	• low output • relatively accurate • needs reference temperature and signal conditioning
strain gauge	Resistive material is stressed and changes resistance.	bridge output	• temperature effect on resistance must be compensated for
load cell	Two strain gauges working in opposition. Two additional resistors are included for temperature compensation.	bridge output	• better temperature compensation than strain gauge • double the sensitivity of a single strain gauge

Example 7.5

An RTD has a resistance of 122 Ω at 20°C. The first order temperature coefficient is $6.00 \times 10^{-3}/°\text{C}$, and the second order temperature coefficient is $9.2 \times 10^{-8}/°\text{C}$. What is the RTD resistance at 566°C?

Solution

$$R_T = R_{T_{\text{ref}}} \left(1 + \alpha_1 \left(T - T_{\text{ref}} \right) + \alpha_2 \left(T - T_{\text{ref}} \right)^2 \right)$$

$$= (122 \ \Omega) \left(\begin{array}{l} 1 + \left(6.00 \times 10^{-3} \ \dfrac{1}{°\text{C}} \right) \\ \times \left(566°\text{C} - 20°\text{C} \right) \\ + \left(9.2 \times 10^{-8} \ \dfrac{1}{°\text{C}^2} \right) \\ \times \left(566°\text{C} - 20°\text{C} \right)^2 \end{array} \right)$$

$$= 525 \ \Omega$$

THERMISTORS

Thermistors are sensitive temperature sensors: they experience large drops in resistance with relatively small increments of increasing temperature, but the relationship between resistance and temperature in thermistors is highly nonlinear. Its characteristics vary between units. A brief description of this sensor can be found in Table 7.2.

Thermistors are semiconductor devices, usually made of silicon. Semiconductors are known for their sensitivity and negative temperature coefficients. There has been some success in developing thermistors that are more linear. Characteristics often vary widely between units.

VARIABLE CAPACITANCE TRANSDUCERS

Most variable capacitive transducers vary their plate spacing as a mechanism for adjusting the capacitance. Characteristics of this type of transducer are listed in Table 7.2.

The equation for capacitance from the capacitance section of Ch. 1 follows. C is the capacitance in farads, A is the capacitor area in square meters, d is the capacitor-plate spacing in meters, ε is the dielectric constant, ε_r is the relative dielectric constant, and ε_0 is the dielectric constant 8.85×10^{-12} F/m for a vacuum dielectric.

$$C = \frac{\varepsilon A}{d} = \frac{\varepsilon_r \varepsilon_0 A}{d}$$

The capacitor holds a charge defined as

$$Q = CV$$

The charge is constant, so the voltage will vary in inverse proportion to the change in capacitance. A capacitive microphone uses this principle to turn sound pressure into electrical signals. Displacement can be measured by changes in the plate spacing.

$$V = \frac{Q}{C}$$

Since the charge is constant,

$$\Delta V = \frac{Q}{\Delta C} \qquad\qquad 7.9$$

The voltage change can be amplified as needed.

Chemical changes can be measured by the change in relative dielectric.

Another use is to vary the frequency of an inductor-capacitor (LC) oscillator, as shown in Fig. 7.2. This can be used as a frequency modulation (FM) transmitter.

Figure 7.2 Capacitive Transducer Using an LC Oscillator

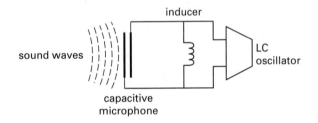

LINEAR VARIABLE DIFFERENTIAL TRANSDUCERS

A *linear variable differential transducer* (LVDT) is a transformer with a movable ferromagnetic core and one primary coil and two secondary coils, as shown in Fig. 7.3. The major characteristics of this type of transducer are listed in Table 7.2.

The center coil is the primary coil, which is driven by an AC source. The outer two coils are secondary coils wound in opposition to each other, so that when the core is exactly centered, the two secondary voltages cancel. As the core moves away from its central position, more voltage is coupled into one secondary coil, and less is coupled into the other. The output voltage is therefore proportional (linearly related) to the core's position. As the core moves from one end through the center, the output voltage changes polarity. Core displacements as small as 2 μm can be resolved.

The displacement is given by the following equation, in which k is the LDVT constant in length (displacement) per volt, and ΔV_{out} is the change in output voltage, V_1 is the initial voltage, and V_2 is the final voltage. The LDVT constant is the reciprocal of the sensitivity.

$$d = k\Delta V_{\text{out}} = k(V_1 - V_2) \qquad 7.10$$

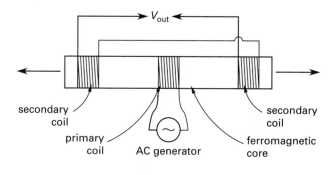

Figure 7.3 Linear Variable Differential Transducer

Example 7.6

An LVDT has a sensitivity of 228 mV/cm. If the initial voltage was 525 mV and the final voltage was 121 mV, how far has the core moved?

Solution

$$d = k\Delta V_{\text{out}} = k(V_1 - V_2)$$
$$= \left(\frac{1 \text{ cm}}{228 \text{ mV}}\right)(525 \text{ mV} - 121 \text{ mV})$$
$$= 1.77 \text{ cm}$$

PIEZOELECTRIC SENSORS

Certain crystals generate a small voltage under mechanical stress. This phenomenon is the *piezoelectric effect*. Quartz, barium titanate, and Rochelle salt are some of the many materials that exhibit this effect. Devices that make use of the effect include microphones, accelerometers, vibration sensors, and electronic oscillator frequency control. Piezoelectric sensors are used to measure engine cylinder pressures and artillery gun barrel pressures. The characteristics of these sensors are described in Table 7.2.

The voltage generated by these crystals is actually a charge similar to that of a capacitor. The crystal charge drains over time as does a capacitor charge. Therefore, measurements of slowly changing quantities would not be accurate. A piezoelectric transducer works best to measure quickly changing stresses.

The piezoelectric effect work both ways. A voltage applied to a crystal causes the crystal to elongate or shorten. The distances the crystal expands or contracts are on the order of micrometers. This effect is used as the foundation for accurate microactuators.

PHOTOSENSITIVE TRANSDUCERS

There are two types of *photosensitive transducers: photovoltaic sensors* generate electricity when exposed to light, and *photoconductive sensors* change their electrical resistance in response to a change in incident light intensity. The characteristics of both types of photosensitive transducers are listed in Table 7.2.

Photovoltaic devices are not only transducers, but also provide power in remote locations away from the power grid. Units for the power generated by photovoltaic transducers are usually watts per square meter (W/m^2). The power output can be calculated with the following equation, in which k is the cell output in W/m^2, A is the area in square meters, and θ is the angle between the cell perpendicular and the sun. If the cell is facing directly into the sun, this angle is $0°$.

$$P_{\text{out}} = kA\cos\theta \qquad 7.11$$

Photoconductive transducers can measure light in the visible, infrared, and ultraviolet spectrums.

Example 7.7

A photovoltaic cell output is 16 W/m^2. The cell's area is 4 cm × 8 cm. Ignoring atmospheric losses, what is the cell's power output if it is directly facing the sun? What is the power if the cell is tilted 30° away from direct sunlight?

Solution

$$P_{\text{out},0°} = kA\cos\theta$$
$$= \left(16 \frac{\text{W}}{\text{m}^2}\right)(0.04 \text{ m})(0.08 \text{ m})\cos 0°$$
$$= 0.051 \text{ W} \quad (51 \text{ mW})$$
$$P_{\text{out},30°} = kA\cos\theta$$
$$= \left(16 \frac{\text{W}}{\text{m}^2}\right)(0.04 \text{ m})(0.08 \text{ m})\cos 30°$$
$$= 0.044 \text{ W} \quad (44 \text{ mW})$$

STRAIN GAUGES

A *strain gauge* monitors the strain (deformation) of an object by measuring a change in the gauge's resistance in response to that strain. The gage material may be a fine wire, metal foil, or a semiconductor. Table 7.2 summarizes the qualities of strain gauges. The resistance equation, introduced in Ch. 1, is

$$R = \frac{\rho l}{A}$$

If the length of the resistor changes by a certain amount, Δl, the resistance change is

$$\Delta R = \frac{\rho \Delta l}{A} \qquad 7.12$$

Since a strain gauge is a resistor, it is subject to resistance changes with temperature. In order to reduce temperature effects, materials with low resistance temperature coefficients are used, as well as temperature compensation. The Wheatstone bridge section discusses some temperature compensation methods. Figure 7.4 shows a typical strain gauge construction. The resistor pulls and lengthens in the sensitive direction. The resistor does not appreciably move in the insensitive direction.

Figure 7.4 Strain Gauge

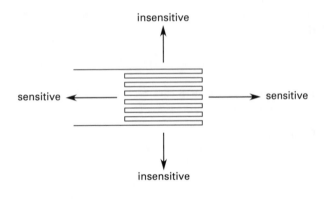

The *gage factor*, GF, or strain sensitivity, is calculated with the following equation, in which $\Delta R/R$ is the fractional resistance change caused by the strain, and $\Delta L/L$ is the fractional length change caused by the strain.

$$GF = \frac{\dfrac{\Delta R}{R}}{\dfrac{\Delta L}{L}} \qquad 7.13$$

Example 7.8

A strain gauge undergoes strain that results in a change in length from 5.00 cm to 5.03 cm. This changes the resistance from 1000.0 Ω to 1013.0 Ω. What is the gage factor?

Solution

$$GF = \frac{\dfrac{\Delta R}{R}}{\dfrac{\Delta L}{L}} = \frac{\dfrac{13.0\ \Omega}{1000\ \Omega}}{\dfrac{0.03\ \text{cm}}{5.00\ \text{cm}}}$$
$$= 2.17$$

Wheatstone Bridge

Strain gauges and *load cells* depend on a change in electrical resistance. Often the resistance changes are small. A *Wheatstone bridge*, an instrument that measures resistance by balancing two legs of a bridge circuit, helps identify these small changes. (Characteristics of the transducers known as load cells can be found in Table 7.2.)

A Wheatstone bridge produces a large current or voltage change in response to a small change in resistance. Figure 7.5 shows a Wheatstone bridge.

Figure 7.5 Wheatstone Bridge

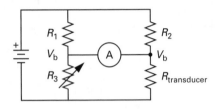

When the bridge is balanced, the current through the meter is 0 A. If the resistance values for R_1 and R_2 are accurately known, the value of $R_{\text{transducer}}$ can be found by adjusting R_3 until the meter reading is 0 A.

$$\frac{R_{\text{transducer}}}{R_2} = \frac{R_3}{R_1} \qquad 7.14$$

Or,

$$R_{\text{transducer}} = \frac{R_2 R_3}{R_1} \qquad 7.15$$

The *bridge constant*, BC, is the ratio of the bridge voltage output change to the transducer resistance change.

$$BC = \frac{\Delta V_{\text{bridge}}}{\Delta R_{\text{transducer}}} \qquad 7.16$$

A dummy or inactive strain gauge can be used as one of the resistors, as shown in Fig. 7.6. Temperature compensation results if the active strain gauge and the dummy strain gauge are at the same temperature. The two should be placed in close proximity to ensure that their temperatures are very nearly identical. A load cell has all four resistors in the same package to enhance temperature compensation.

Example 7.9

The bridge circuit in Fig. 7.6 has resistance values of $R_1 = R_2 = R_{\text{dummy gauge}} = 1.000$ kΩ. The active gauge has an unstressed resistance of 1.000 kΩ, and a gage factor of 2.2. The voltage source is 20 V. A strain of 1100 μm/m is applied. What is the bridge offset voltage?

Figure 7.6 Wheatstone Bridge with Strain Gauge

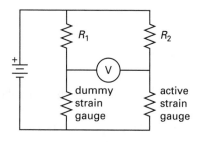

Solution

$$\text{GF} = \frac{\dfrac{\Delta R}{R}}{\text{strain}}$$

$$\begin{aligned}
\Delta R &= (\text{GF}) \, (\text{strain}) \, R \\
&= (2.2) \left(1.1 \times 10^{-3} \; \frac{\text{m}}{\text{m}} \right) (1.000 \; \text{k}\Omega) \left(1000 \; \frac{\Omega}{\text{k}\Omega} \right) \\
&= 2.42 \; \Omega
\end{aligned}$$

The strain gauge resistance is 1002.42 Ω. The resistance of R_2 is given in the problem as 1.000 kΩ (1000 Ω). The voltage at the junction of R_2 and the active strain gauge is

$$\begin{aligned}
V_{\text{out}} &= V_{\text{source}} \left(\frac{R_{\text{gauge}}}{R_2 + R_{\text{gauge}}} \right) \\
&= (20 \; \text{V}) \left(\frac{1002.42 \; \Omega}{1000 \; \Omega + 1002.42 \; \Omega} \right) \\
&= 10.0121 \; \text{V}
\end{aligned}$$

The offset voltage is

$$10.0121 \; \text{V} - 10.000 \; \text{V} = 0.0121 \; \text{V} \quad (12.1 \; \text{mV})$$

LASER INSTRUMENTS

Lasers are increasingly used in instrumentation. One use is determining distances, or ranging. Laser imaging detection and ranging devices (lidar), also called light detection and ranging devices, fire laser pulses at a reflector and record the time it takes the pulses to return. The speed of light, c, is a constant 2.998×10^8 m/s, so the distance, d, measured by a lidar pulse in a certain time, t, is

$$d = \frac{ct}{2} \qquad \qquad 7.17$$

The laser pulse travels the distance twice as it goes to the reflector and back, so the product of the speed of light and the time must be divided by two.

Example 7.10

A laser rangefinder takes 2.203 μs to receive its transmitted pulse from a reflector. What is the distance to the reflector?

Solution

$$\begin{aligned}
d &= \frac{ct}{2} \\
&= \frac{\left(2.998 \times 10^8 \; \dfrac{\text{m}}{\text{s}} \right) \left(2.203 \times 10^{-6} \; \text{s} \right)}{2} \\
&= 330.23 \; \text{m}
\end{aligned}$$

PRACTICE PROBLEMS

1. A 10-bit ADC measures voltages between -5.0 V and 5.0 V. What is most nearly its resolution?

(A) 1 nV/step
(B) 5 mV/step
(C) 10 mV/step
(D) 1 V/step

2. A 10-bit ADC has an input ranging between -10 V and 10 V. What is most nearly the conversion gain?

(A) 0.001 step/V
(B) 1 step/V
(C) 50 steps/V
(D) 100 steps/V

3. An ammeter reads 2.01 A. The true value is 1.98 A. What is most nearly the percentage error?

(A) -1.52%
(B) -1.49%
(C) -0.0152%
(D) 1.52%

4. A type K thermocouple measures 600°F. Its reference junction is at the melting point of ice. Its thermocouple constant is 21 μV/°F. What is most nearly the voltage across the thermocouple?

(A) 11.9 mV
(B) 12.6 mV
(C) 22.6 mV
(D) 22.8 mV

5. An RTD has a resistance of 550 Ω at 25°C. The first order temperature coefficient is 5.80×10^{-3} 1/°C. The second order temperature coefficient is 8.1×10^{-8} 1/°C. What is most nearly the RTD resistance at 777°C?

(A) 2940 Ω
(B) 2950 Ω
(C) 2970 Ω
(D) 3050 Ω

6. An LVDT has a sensitivity of 25.6 mV/mm. The initial voltage was 541 mV. After the core moved, the final voltage is −336 mV. What is most nearly the distance the core moved?

(A) −13 mm
(B) 0.29 mm
(C) 8.0 mm
(D) 34 mm

7. A photoelectric (photovoltaic) cell is rated at 19.9 W/m² under full sunlight. Assume the sunlight is 82% of full brightness. The cell is at an angle of 22° relative to the sun. The cell's dimensions are 12 cm × 24 cm. What is most nearly the output of the cell?

(A) 440 mW
(B) 470 mW
(C) 530 mW
(D) 4400 W

8. A strain gauge changes its length from 44.0 mm to 44.3 mm under load. The resistance changes from 220 Ω to 225 Ω. What is most nearly the gage factor?

(A) 0.023
(B) 0.30
(C) 3.3
(D) 17

9. The Wheatstone bridge in the illustration has a strain gauge in one leg. The excitation voltage is 12.00 V, R_1 and R_2 are 1.000 kΩ each, the dummy strain gauge is 1.000 kΩ, and the nominal strain gauge resistance is 1000 Ω. A stress is put onto the strain gauge increasing its resistance to 1010 Ω. What is the voltage change at the junction of R_2 and the strain gauge?

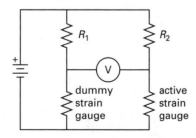

(A) 0.0 mV
(B) 15 mV
(C) 30 mV
(D) 6000 mV

10. A laser rangefinder sends a pulse to a reflector. The time for the pulse to return to the rangefinder is 4.302 μs. What is most nearly the distance to the reflector?

(A) 6.5 m
(B) 650 m
(C) 1300 m
(D) 2600 m

SOLUTIONS

1. The ADC resolution is

$$\text{resolution} = \frac{\text{input range}}{2^n \text{ steps}}$$

$$= \frac{5.0 \text{ V} - (-5.0 \text{ V})}{(2)^{10} \text{ steps}}$$

$$= 0.009\,765\,6 \text{ V/step} \quad (10 \text{ mV/step})$$

Note: Option (A) uses base 10 instead of base 2. Option (B) is half the range divided by $(2)^{10}$. Option (D) is the range divided by number of bits.

Answer is C.

2. The conversion gain is

$$A_{\text{ADC}} = \frac{\text{output}}{\text{input}}$$

$$= \frac{(2)^{10} \text{ steps}}{10 \text{ V} - (-10 \text{ V})}$$

$$= 51.2 \text{ steps/V} \quad (50 \text{ steps/V})$$

Note: Option (A) is the inverse of ADC gain, and could represent the gain for a DAC. Option (B) is the peak positive voltage over the bits, which is not the definition of conversion gain. Option (D) uses 10 V as the range, when the correct range should be 20 V (−10 V to 10 V).

Answer is C.

3. The percentage error is

$$E = \left(\frac{\text{measured value} - \text{true value}}{\text{true value}} \right) \times 100\%$$

$$= \left(\frac{2.01 \text{ A} - 1.98 \text{ A}}{1.98 \text{ A}} \right) \times 100\%$$

$$= 1.515\% \quad (1.52\%)$$

Note: In option (A), the measured value is subtracted from the correct value. This is the reverse of the correct method. Option (B) uses the measured value in the denominator, not the true value. Also the measured valued is subtracted from the true value. Option (C) does not multiply by 100%.

Answer is D.

4. The voltage across the thermocouple can be calculated as follows.

$$V = k_T \left(T - T_{\text{ref}} \right)$$

$$= \left(21 \frac{\mu\text{V}}{°\text{F}} \right) (600°\text{F} - 32°\text{F})$$

$$= 11\,928 \ \mu\text{V} \quad (11.9 \text{ mV})$$

Note: In option (B), the reference temperature is missing. In option (C), the thermocouple constant for Celsius is used instead of the one for Fahrenheit. In option (D), the thermocouple constant for Celsius is used instead of the one for Fahrenheit, and the wrong reference temperature is used.

Answer is A.

5. The RTD resistance at 777°C is

$$R_T = R_{T_{\text{ref}}} \left(1 + \alpha_1 \left(T - T_{\text{ref}} \right) + \alpha_2 \left(T - T_{\text{ref}} \right)^2 \right)$$

$$= R_T \left(1 + \alpha_1 \left(\Delta T \right) + \alpha_2 \left(\Delta T \right)^2 \right)$$

$$= (550 \ \Omega) \begin{pmatrix} 1 + \left(5.80 \times 10^{-3} \ \dfrac{1}{°\text{C}} \right) \\ \times (777°\text{C} - 25°\text{C}) \\ + \left(8.1 \times 10^{-8} \ \dfrac{1}{°\text{C}} \right) \\ \times (777°\text{C} - 25°\text{C})^2 \end{pmatrix}$$

$$= 2974 \ \Omega \quad (2970 \ \Omega)$$

Note: Option (A) is missing the second order temperature term. In Option (B), the second order term is not squared. Option (D) is missing the 25°C term.

Answer is C.

6. The distance traveled by the core is

$$d = k(V_1 - V_2)$$

$$= \left(\frac{1 \text{ mm}}{25.6 \text{ mV}} \right) (541 \text{ mV} - (-336 \text{ mV}))$$

$$= 34.258 \text{ mm} \quad (34 \text{ mm})$$

Note: Option (A) does not use the initial voltage. In option (B), the variables are inverted. Option (C) subtracts the final voltage instead of adding it.

Answer is D.

7. The power output of the cell is

$$P_{\text{out}} = kA \cos \theta$$

$$= \left(19.9 \ \frac{\text{W}}{\text{m}^2} \right) (0.12 \text{ m}) (0.24 \text{ m}) (0.82) \cos 22°$$

$$= 0.436 \text{ W} \quad (440 \text{ mW})$$

Note: Option (B) does not factor into the equation the cell's angle relative to the sun. Option (C) does not factor the reduced sunlight into the equation. In option (D), the cell dimensions are incorrect: the decimal point is wrong.

Answer is A.

8. The gage factor is

$$\text{GF} = \frac{\dfrac{\Delta R}{R}}{\dfrac{\Delta L}{L}} = \frac{\dfrac{5 \ \Omega}{220 \ \Omega}}{\dfrac{0.3 \text{ mm}}{44.0 \text{ mm}}}$$

$$= 3.333 \quad (3.3)$$

Note: Option (A) left the lengths out of the equation. Option (B) inverts the resistances and lengths in the equation. In option (D), only the changes in length and resistance are factored into the equation: the initial values are neglected.

Answer is C.

9. Calculate the voltage with no load and the voltage with load to find the voltage change.

$$V_{\text{out,no load}} = V_{\text{source}} \left(\frac{R_{\text{gauge}}}{R_2 + R_{\text{gauge}}} \right)$$

$$= (12.000 \text{ V}) \left(\frac{1000 \ \Omega}{1000 \ \Omega + 1000 \ \Omega} \right)$$

$$= 6.000 \text{ V}$$

$$V_{\text{out,load}} = V_{\text{source}} \left(\frac{R_{\text{gauge}}}{R_2 + R_{\text{gauge}}} \right)$$

$$= (12.000 \text{ V}) \left(\frac{1010 \ \Omega}{1000 \ \Omega + 1010 \ \Omega} \right)$$

$$= 6.030 \text{ V}$$

The voltage change is

$$6.030 \text{ V} - 6.000 \text{ V} = 0.030 \text{ V} \quad (30 \text{ mV})$$

Note: Option (A) misuses the equation and shows the resistance of R_2 and the strain gauge changing the same amount, which results in no voltage change. In option (B), the voltage of 6.000 V at the junction of R_1 and the dummy strain gauge is used instead of the supply voltage of 12.000 V. In option (D), the R_{gauge} resistance is not put into the numerator: the resistance of R_2 is mistakenly used instead.

Answer is C.

10. The distance between the rangefinder and the reflector is

$$d = \frac{ct}{2}$$

$$= \frac{\left(2.998 \times 10^8 \ \frac{\text{m}}{\text{s}} \right) \left(4.302 \times 10^{-6} \ \text{s} \right)}{2}$$

$$= 644.87 \text{ m} \quad (650 \text{ m})$$

Note: Exponents that describe the SI units are left out of the calculation for option (A). Option (C) fails to divide by two to get the one-way distance and so ends up with the distance for the round trip. Option (D) also fails to divide by two, and instead multiplies by two.

Answer is B.

Index